ECAA
Practice Papers

Copyright © 2018 *UniAdmissions*. All rights reserved.

ISBN 978-1-912557-19-6

No part of this publication may be reproduced or transmitted in any form or by any means, electronic or mechanical, including photocopying, recording, or by any information retrieval system without prior written permission of the publisher. This publication may not be used in conjunction with or to support any commercial undertaking without the prior written permission of the publisher.

Published by *RAR Medical Services Limited*
www.uniadmissions.co.uk
info@uniadmissions.co.uk
Tel: 0208 068 0438

This book is neither created nor endorsed by ECAA. The authors and publisher are not affiliated with ECAA. The information offered in this book is purely advisory and any advice given should be taken within this context. As such, the publishers and authors accept no liability whatsoever for the outcome of any applicant's ECAA performance, the outcome of any university applications or for any other loss. Although every precaution has been taken in the preparation of this book, the publisher and author assume no responsibility for errors or omissions of any kind. Neither is any liability assumed for damages resulting from the use of information contained herein. This does not affect your statutory rights.

ECAA Practice Papers

2 Full Papers & Solutions

David Meacham
Dr Rohan Agarwal

About the Authors

David is a **Merger & Acquisitions Associate** at The Hut Group, a leading online retailer and brand owner in the Beauty & Wellness sectors. Prior to joining The Hut Group, he worked in roles at the Professional Service firm Deloitte, the Investment Bank Greenhill and the Private Equity firm Hgcapital.

David graduated with a **first class honours** in Economics from Gonville and Caius College Cambridge, where he received two college scholarships for outstanding academic performance, in addition to an Essay Prize. He is also a qualified accountant and chartered tax adviser, passing all exams first-time with multiple regional top scores. Since graduating, David has tutored & successfully provided academic coaching to hundreds of students, both in a personal capacity and for university admissions.

Rohan is the **Director of Operations** at *UniAdmissions* and is responsible for its technical and commercial arms. He graduated from Gonville and Caius College, Cambridge and is a fully qualified doctor. Over the last five years, he has tutored hundreds of successful Oxbridge and Medical applicants. He has also authored thirty books on admissions tests and interviews.

Rohan has taught physiology to undergraduates and interviewed medical school applicants for Cambridge. He has published research on bone physiology and writes education articles for the Independent and Huffington Post. In his spare time, Rohan enjoys playing the piano and table tennis.

INTRODUCTION 6

- GENERAL ADVICE 7
- REVISION TIMETABLE 12
- GETTING THE MOST OUT OF MOCK PAPERS 12
- BEFORE USING THIS BOOK 13
- SECTION 1A: AN OVERVIEW 16
- SECTION 1B: AN OVERVIEW 17
- SECTION 2: AN OVERVIEW 18
- MATHS REVISION CHECKLIST 21
- HOW TO USE THIS BOOK 22
- SCORING TABLES 23

MOCK PAPER A 24

- SECTION 1A 24
- SECTION 1B 30
- SECTION 2 33

MOCK PAPER B 35

- SECTION 1A 35
- SECTION 1B 42
- SECTION 2 45

ANSWER KEY 48

MOCK PAPER A ANSWERS 49

- SECTION 1A 49
- SECTION 1B 51
- SECTION 2 54

MOCK PAPER B ANSWERS 56

- SECTION 1A 56
- SECTION 1B 59
- SECTION 2 63

FINAL ADVICE 65

YOUR FREE BOOKS 67

ECAA INTENSIVE COURSE 68

OXBRIDGE INTERVIEW COURSE 69

Introduction

The Basics

The Economics Admissions Assessment (ECAA) is the 2-hour written aptitude exam taken by students applying for economics courses at the most competitive universities.

It is a highly time pressured exam that forces you to apply GCSE and A-level knowledge in ways you have never thought about before. In this respect simply remembering solutions taught in class or from past papers is not enough.

However, fear not, despite what people say, you can actually prepare for the ECAA! With a little practice you can train your brain to manipulate and apply learnt methodologies to novel problems with ease. The best way to do this is through exposure to as many past/specimen papers as you can.

Preparing for the ECAA

Before going any further, it's important that you understand the optimal way to prepare for the ECAA. Rather than jumping straight into doing mock papers, it's essential that you start by understanding the components and the theory behind the ECAA by using an ECAA textbook. Once you've finished the non-timed practice questions, you can progress to past ECAA papers. These are freely available online at **www.uniadmissions.co.uk/ecaa-past-papers** and serve as excellent practice. You're strongly advised to use these in combination with the *ECAA Past Worked Solutions* Book so that you can improve your weaknesses. Finally, once you've exhausted past papers, move onto the mock papers in this book.

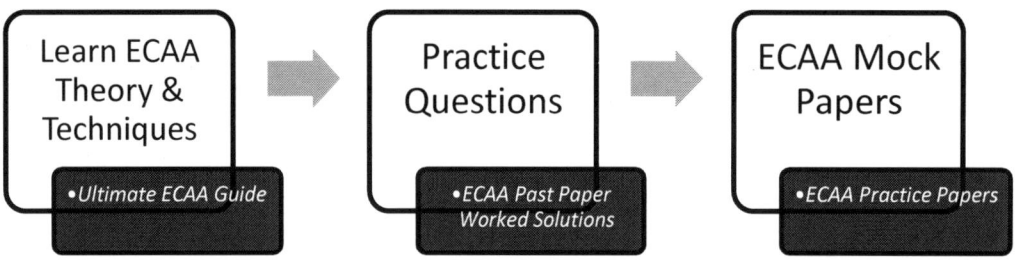

Already seen them all?

So, you've run out of past papers? Well that is where this book comes in. This book contains two unique mock papers; each compiled by Cambridge economics tutors at *UniAdmissions* and available nowhere else.

Having successfully gained a place on their course of choice, our tutors are intimately familiar with the ECAA and its associated admission procedures. So, the novel questions presented to you here are of the correct style and difficulty to continue your revision and stretch you to meet the demands of the ECAA.

General Advice

Start Early
It is much easier to prepare if you practice little and often. Start your preparation well in advance; ideally 10 weeks but at the latest within a month. This way you will have plenty of time to complete as many papers as you wish to feel comfortable and won't have to panic and cram just before the test, which is a much less effective and more stressful way to learn. In general, an early start will give you the opportunity to identify the complex issues and work at your own pace.

Prioritise
Some questions in sections can be long and complex – and given the intense time pressure you need to know your limits. It is essential that you don't get stuck with very difficult questions. If a question looks particularly long or complex, mark it for review and move on. You don't want to be caught 5 questions short at the end just because you took more than 3 minutes in answering a challenging multi-step question. If a question is taking too long, choose a sensible answer and move on. Remember that each question carries equal weighting and therefore, you should adjust your timing in accordingly. With practice and discipline, you can get very good at this and learn to maximise your efficiency.

Positive Marking
There are no penalties for incorrect answers; you will gain one for each right answer and will not get one for each wrong or unanswered one. This provides you with the luxury that you can always guess should you absolutely be not able to figure out the right answer for a question or run behind time. Since each question provides you with 4 to 6 possible answers, you have a 16-25% chance of guessing correctly. Therefore, if you aren't sure (and are running short of time), then make an educated guess and move on. Before 'guessing' you should try to eliminate a couple of answers to increase your chances of getting the question correct. For example, if a question has 5 options and you manage to eliminate 2 options- your chances of getting the question increase from 20% to 33%!

Avoid losing easy marks on other questions because of poor exam technique. Similarly, if you have failed to finish the exam, take the last 10 seconds to guess the remaining questions to at least give yourself a chance of getting them right.

Practice
This is the best way of familiarising yourself with the style of questions and the timing for this section. Although the exam will essentially only test GCSE level knowledge, you are unlikely to be familiar with the style of questions in all sections when you first encounter them. Therefore, you want to be comfortable at using this before you sit the test.

Practising questions will put you at ease and make you more comfortable with the exam. The more comfortable you are, the less you will panic on the test day and the more likely you are to score highly. Initially, work through the questions at your own pace, and spend time carefully reading the questions and looking at any additional data. When it becomes closer to the test, **make sure you practice the questions under exam conditions**.

Past Papers

Official past papers and answers from 2016 onwards are freely available online on our website at www.uniadmissions.co.uk/product/ecaa-mock-papers.

You will undoubtedly get stuck when doing some past paper questions – they are designed to be tricky and the answer schemes don't offer any explanations. Thus, **you're highly advised to acquire a copy of ECAA *Past Paper Worked Solutions*** – a free ebook is available online (see the back of this book for more details).

Repeat Questions

When checking through answers, pay particular attention to questions you have got wrong. If there is a worked answer, look through that carefully until you feel confident that you understand the reasoning, and then repeat the question without help to check that you can do it. If only the answer is given, have another look at the question and try to work out why that answer is correct. This is the best way to learn from your mistakes, and means you are less likely to make similar mistakes when it comes to the test. The same applies for questions which you were unsure of and made an educated guess which was correct, even if you got it right. When working through this book, **make sure you highlight any questions you are unsure of**, this means you know to spend more time looking over them once marked.

No Calculators

You aren't permitted to use calculators in the exam – thus, it is essential that you have strong numerical skills. For instance, you should be able to rapidly convert between percentages, decimals and fractions. You will seldom get questions that would require calculators, but you would be expected to be able to arrive at a sensible estimate. Consider for example:

Estimate 3.962 x 2.322;

3.962 is approximately 4 and 2.323 is approximately 2.33 = 7/3.

Thus, $3.962 x 2.322 4 x \frac{7}{3} = \frac{28}{3} = 9.33$

Since you will rarely be asked to perform difficult calculations, you can use this as a signpost of if you are tackling a question correctly. For example, when solving a physics question, you end up having to divide 8,079 by 357- this should raise alarm bells as calculations in the ECAA are rarely this difficult.

> ***Top tip!*** In general, students tend to improve the fastest in section 2 and slowest in section 1; section 3 usually falls somewhere in the middle. Thus, if you have very little time left, it's best to prioritise section 2.

A word on timing...

"If you had all day to do your exam, you would get 100%. But you don't."
Whilst this isn't completely true, it illustrates a very important point. Once you've practiced and know how to answer the questions, the clock is your biggest enemy. This seemingly obvious statement has one very important consequence. **The way to improve your score is to improve your speed.** There is no magic bullet. But there are a great number of techniques that, with practice, will give you significant time gains, allowing you to answer more questions and score more marks.

Timing is tight throughout – **mastering timing is the first key to success**. Some candidates choose to work as quickly as possible to save up time at the end to check back, but this is generally not the best way to do it. Often questions can have a lot of information in them – each time you start answering a question it takes time to get familiar with the instructions and information. By splitting the question into two sessions (the first run-through and the return-to-check) you double the amount of time you spend on familiarising yourself with the data, as you have to do it twice instead of only once. This costs valuable time. In addition, candidates who do check back may spend 2–3 minutes doing so and yet not make any actual changes. Whilst this can be reassuring, it is a false reassurance as it is unlikely to have a significant effect on your actual score. Therefore, it is usually best to pace yourself very steadily, aiming to spend the same amount of time on each question and finish the final question in a section just as time runs out. This reduces the time spent on re-familiarising with questions and maximises the time spent on the first attempt, gaining more marks.

It is essential that you don't get stuck with the hardest questions – no doubt there will be some. In the time spent answering only one of these you may miss out on answering three easier questions. If a question is taking too long, choose a sensible answer and move on. Never see this as giving up or in any way failing, rather it is the smart way to approach a test with a tight time limit. With practice and discipline, you can get very good at this and learn to maximise your efficiency. It is not about being a hero and aiming for full marks – this is almost impossible and very much unnecessary. It is about maximising your efficiency and gaining the highest possible number of marks within the time you have.

Use the Options:
Some questions may try to overload you with information. When presented with large tables and data, it's essential you look at the answer options so you can focus your mind. This can allow you to reach the correct answer a lot more quickly. Consider the example below:

The table below shows the results of a study investigating antibiotic resistance in staphylococcus populations. A single staphylococcus bacterium is chosen at random from a similar population. Resistance to any one antibiotic is independent of resistance to others.

Calculate the probability that the bacterium selected will be resistant to all four drugs.

A. 1 in 10^6
B. 1 in 10^{12}
C. 1 in 10^{20}
D. 1 in 10^{25}
E. 1 in 10^{30}
F. 1 in 10^{35}

Antibiotic	Number of Bacteria tested	Number of Resistant Bacteria
Benzyl-penicillin	10^{11}	98
Chloramphenicol	10^9	1200
Metronidazole	10^8	256
Erythromycin	10^5	2

Looking at the options first makes it obvious that there is **no need to calculate exact values**- only in powers of 10. This makes your life a lot easier. If you hadn't noticed this, you might have spent well over 90 seconds trying to calculate the exact value when it wasn't even being asked for.

In other cases, you may actually be able to use the options to arrive at the solution quicker than if you had tried to solve the question as you normally would. Consider the example below:

A region is defined by the two inequalities: $x - y^2 > 1 \wedge xy > 1$. Which of the following points is in the defined region?

A. (10,3)
B. (10,2)
C. (-10,3)
D. (-10,2)
E. (-10,-3)

Whilst it's possible to solve this question both algebraically or graphically by manipulating the identities, by far **the quickest way is to actually use the options**. Note that options C, D and E violate the second inequality, narrowing down to answer to either A. or B. For A: $10 - 3^2 = 1$ and thus this point is on the boundary of the defined region and not actually in the region. Thus the answer is B (as $10-4 = 6 > 1$.)

In general, it pays dividends to look at the options briefly and see if they can be help you arrive at the question more quickly. Get into this habit early – it may feel unnatural at first but it's guaranteed to save you **time in the long run**.

Keywords
If you're stuck on a question; pay particular attention to the options that contain key modifiers like "**always**", "**only**", "**all**" as examiners like using them to test if there are any gaps in your knowledge. E.g. the statement "arteries carry oxygenated blood" would normally be true; "All arteries carry oxygenated blood" would be false because the pulmonary artery carries deoxygenated blood.

Manage your Time:

It is highly likely that you will be juggling your revision alongside your normal school studies. Whilst it is tempting to put your A-levels on the back burner falling behind in your school subjects is not a good idea, don't forget that to meet the conditions of your offer should you get one you will need at least one A*. So, time management is key!

Make sure you set aside a dedicated 90 minutes (and much more closer to the exam) to commit to your revision each day. The key here is not to sacrifice too many of your extracurricular activities, everybody needs some down time, but instead to be efficient. Take a look at our list of top tips for increasing revision efficiency below:

1. Create a comfortable work station: Declutter and stay tidy
2. Treat yourself to some nice stationery
3. See if music works for you – if not, find somewhere peaceful and quiet to work
4. Turn off your mobile or at least put it into silent mode and silence social media alerts
5. Keep the TV off and out of sight
6. Stay organised with to do lists and revision timetables – more importantly, stick to them!
7. Keep to your set study times and don't bite off more than you can chew
8. Study while you're commuting
9. Adopt a positive mental attitude
10. Get into a routine
11. Consider forming a study group to focus on the harder exam concepts
12. Plan rest and reward days into your timetable – these are excellent incentive for you to stay on track with your study plans!

Keep Fit & Eat Well:

'A car won't work if you fill it with the wrong fuel' - your body is exactly the same. You cannot hope to perform unless you remain fit and well. The best way to do this is not underestimate the importance of healthy eating. Beige, starchy foods will make you sluggish; instead start the day with a hearty breakfast like porridge. Aim for the recommended 'five a day' intake of fruit/veg and stock up on the oily fish or blueberries – the so called "super foods".

When hitting the books, it's essential to keep your brain hydrated. If you get dehydrated you'll find yourself lethargic and possibly developing a headache, neither of which will do any favours for your revision. Invest in a good water bottle that you know the total volume of and keep sipping throughout the day. Don't forget that the amount of water you should be aiming to drink varies depending on your mass, so calculate your own personal recommended intake as follows: 30 ml per kg per day.

It is well known that exercise boosts your wellbeing and instils a sense of discipline. All of which will reflect well in your revision. It's well worth devoting half an hour a day to some exercise, get your heart rate up, break a sweat, and get those endorphins flowing.

Sleep

It's no secret that when revising you need to keep well rested. Don't be tempted to stay up late revising as sleep actually plays an important part in consolidating long term memory. Instead aim for a minimum of 7 hours good sleep each night, in a dark room without any glow from electronic appliances. Install flux (https://justgetflux.com) on your laptop to prevent your computer from disrupting your circadian rhythm. Aim to go to bed the same time each night and no hitting snooze on the alarm clock in the morning!

Revision Timetable

Still struggling to get organised? Then try filling in the example revision timetable below, remember to factor in enough time for short breaks, and stick to it! Remember to schedule in several breaks throughout the day and actually use them to do something you enjoy e.g. TV, reading, YouTube etc.

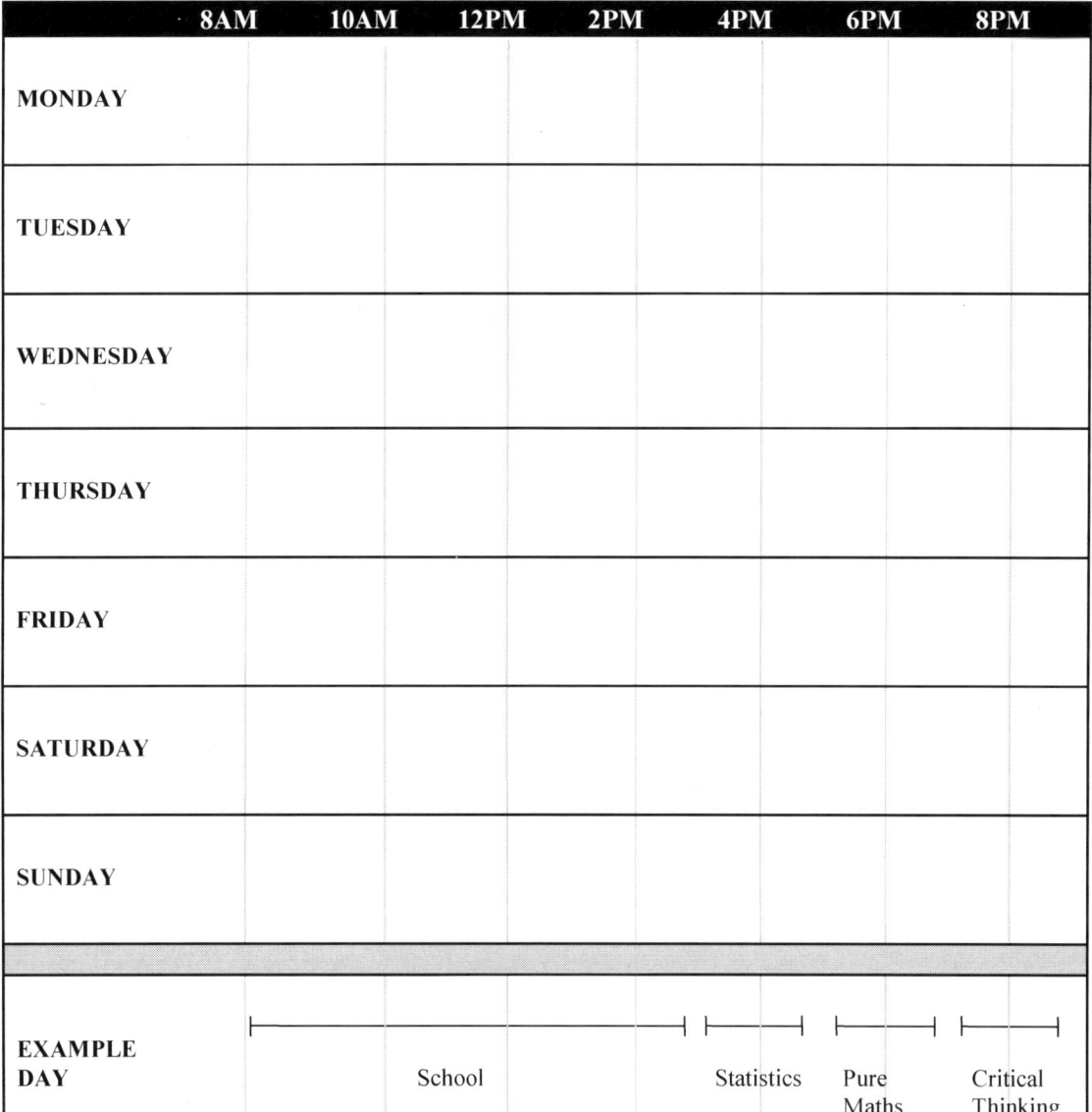

Top tip! Ensure that you take a watch that can show you the time in seconds into the exam. This will allow you have a much more accurate idea of the time you're spending on a question. In general, if you've spent >150 seconds on a section 1 question – move on regardless of how close you think you are to solving it.

Getting the most out of Mock Papers

Mock exams can prove invaluable if tackled correctly. Not only do they encourage you to start revision earlier, they also allow you to **practice and perfect your revision technique**. They are often the best way of improving your knowledge base or reinforcing what you have learnt. Probably the best reason for attempting mock papers is to familiarise yourself with the exam conditions of the ECAA as they are particularly tough.

Start Revision Earlier
Thirty five percent of students agree that they procrastinate to a degree that is detrimental to their exam performance. This is partly explained by the fact that they often seem a long way in the future. In the scientific literature this is well recognised, Dr. Piers Steel, an expert on the field of motivation states that *'the further away an event is, the less impact it has on your decisions'*.

Mock exams are therefore a way of giving you a target to work towards and motivate you in the run up to the real thing – every time you do one treat it as the real deal! If you do well then it's a reassuring sign; if you do poorly then it will motivate you to work harder (and earlier!).

Practice and perfect revision techniques
In case you haven't realised already, revision is a skill all to itself, and can take some time to learn. For example, the most common revision techniques including **highlighting and/or re-reading are quite ineffective** ways of committing things to memory. Unless you are thinking critically about something you are much less likely to remember it or indeed understand it.

Mock exams, therefore allow you to test your revision strategies as you go along. Try spacing out your revision sessions so you have time to forget what you have learnt in-between. This may sound counterintuitive but the second time you remember it for longer. Try teaching another student what you have learnt, this forces you to structure the information in a logical way that may aid memory. Always try to question what you have learnt and appraise its validity. Not only does this aid memory but it is also a useful skill for Oxbridge interviews and beyond.

Improve your knowledge
The act of applying what you have learnt reinforces that piece of knowledge. A question may ask you to think about a relatively basic concept in a novel way (not cited in textbooks), and so deepen your understanding. Exams rarely test word for word what is in the syllabus, so when running through mock papers try to understand how the basic facts are applied and tested in the exam. As you go through the mocks or past papers take note of your performance and see if you consistently under-perform in specific areas, thus highlighting areas for future study.

Get familiar with exam conditions
Pressure can cause all sorts of trouble for even the most brilliant students. The ECAA is a particularly time pressured exam with high stakes – your future (without exaggerating) does depend on your result to a great extent. The real key to the ECAA is overcoming this pressure and remaining calm to allow you to think efficiently.

Mock exams are therefore an excellent opportunity to devise and perfect your own exam techniques to beat the pressure and meet the demands of the exam. **Don't treat mock exams like practice questions – it's imperative you do them under time conditions.**

> *Remember!* It's better that you make all the mistakes you possibly can now in mock papers and then learn from them so as not to repeat them in the real exam.

Before using this Book

Do the ground work
- Read in detail: the background, methods, and aims of the ECAA as well logistical considerations such as how to take the ECAA in practice. A good place to start is a ECAA textbook like *The Ultimate ECAA Guide* (flick to the back to get a free copy!) which covers all the groundwork.
    - It is generally a good idea to start re-capping all your GCSE and AS maths.
- Remember that calculators are not permitted in the exam, so get comfortable doing more complex long addition, multiplication, division, and subtraction.
- Get comfortable rapidly converting between percentages, decimals, and fractions.

- Practice developing logical arguments and structuring essays with an obvious introduction, main body, and ending.
- These are all things which are easiest to do alongside your revision for exams before the summer break. Not only gaining a head start on your ECAA revision but also complimenting your year 12 studies well.
- Discuss topical economics problems with others - propose theories and be ready to defend your argument. This will rapidly build your scientific understanding for section 2 but also prepare you well for an oxbridge interview.
- Read through the ECAA syllabus before you start tackling whole papers. This is absolutely essential. It contains several stated formulae, constants, and facts that you are expected to apply - or may just be an answer in their own right. Familiarising yourself with the syllabus is also a quick way of teaching yourself the additional information other exam boards may learn which you do not. Sifting through the whole ECAA syllabus is a time-consuming process so we have done it for you. **Be sure to flick through the syllabus checklist** later on, which also doubles up as a great revision aid for the night before!

Ease in gently

With the ground work laid, there's still no point in adopting exam conditions straight away. Instead invest in a beginner's guide to the ECAA, which will not only describe in detail the background and theory of the exam, but take you through section by section what is expected. *The Ultimate ECAA Guide* is the most popular ECAA textbook – you can get a free copy by flicking to the back of this book.

When you are ready to move on to past papers, take your time and puzzle your way through all the questions. Really try to understand solutions. A past paper question won't be repeated in your real exam, so don't rote learn methods or facts. Instead, focus on applying prior knowledge to formulate your own approach.

If you're really struggling and have to take a sneak peek at the answers, then practice thinking of alternative solutions, or arguments for essays. It is unlikely that your answer will be more elegant or succinct than the model answer, but it is still a good task for encouraging creativity with your thinking. Get used to thinking outside the box!

Accelerate and Intensify

Start adopting exam conditions after you've done two past papers. Don't forget that **it's the time pressure that makes the ECAA hard** – if you had as long as you wanted to sit the exam you would probably get 100%. If you're struggling to find comprehensive answers to past papers then ECAA *Past Papers Worked Solutions* contains detailed explained answers to every ECAA past paper question and essay (flick to the back to get a free copy).

Doing every past paper at least twice is a good target for your revision. In any case, choose a paper and proceed with strict exam conditions. Take a short break and then mark your answers before reviewing your progress. For revision purposes, as you go along, keep track of those questions that you guess – these are equally as important to review as those you get wrong.

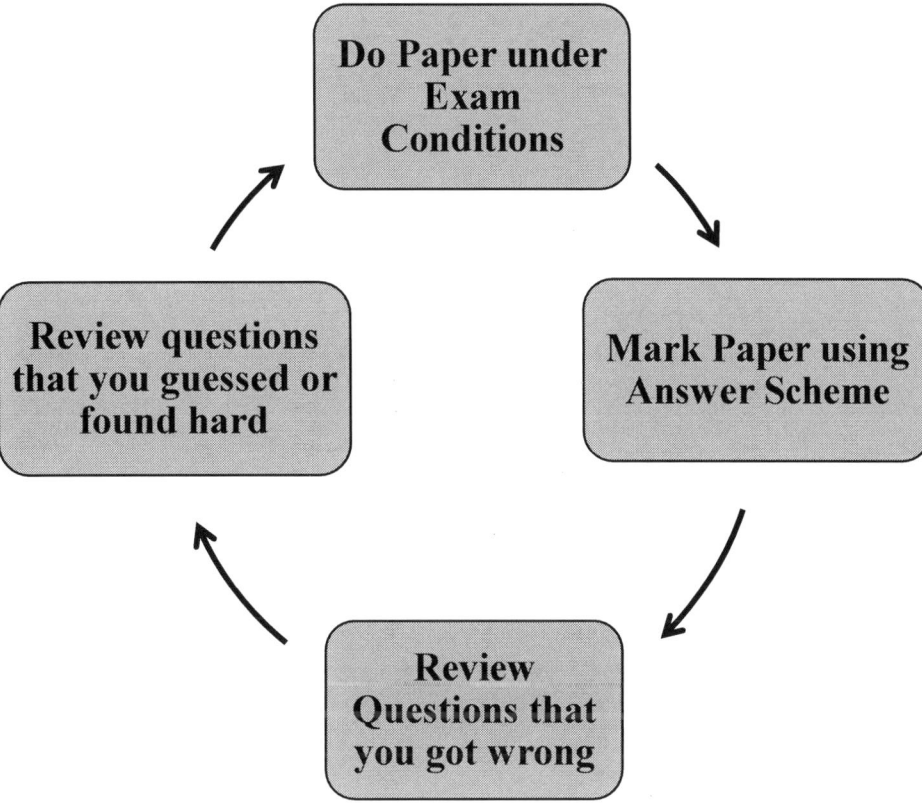

Once you've exhausted all the past papers, move on to tackling the unique mock papers in this book. In general, you should aim to complete one to two mock papers every night in the ten days preceding your exam.

Section 1A: An Overview

What will you be tested on?	No. of Questions	Duration
Problem-solving skills, numerical and spatial reasoning, critical thinking skills, understanding arguments and reasoning	20 MCQs	80 Minutes (incl. Section 1B)

This is the first section of the ECAA, comprising a total of 20 MCQ questions. You have 90 minutes in total to complete section 1, including the maths part (section 1B). It's best to devote 40 minutes to each subsection which gives you approximately 2 minutes for each question.

Not all the questions are of equal difficulty and so as you work through the past material it is certainly worth learning to recognise quickly which questions you should spend less time on in order to give yourself more time for the trickier questions.

Deducing arguments
Several MCQ questions will be aimed at testing your understand of the writer's argument. It is common to see questions asking you 'what is the writer's view?' or 'what is the writer trying to argue?'. This is arguably an important skill you will have to develop, and the TSA is designed to test this ability. You have limited time to read the passage and understand the writer's argument, and the only way to improve your reading comprehension skill is to read several well-written news articles on a daily basis and think about them in a critical manner.

Assumptions
It is important to be able to identify the assumptions that a writer makes in the passage, as several questions might question your understanding of what is assumed in the passage. For example, if a writer mentions that 'if all else remains the same, we can expect our economic growth to improve next year', you can identify an assumption being made here – the writer is clearly assuming that all external factors remain the same.

Fact vs. Opinion
It is important to **be able to decipher whether the writer is stating a fact or an opinion** – the distinction is usually rather subtle and you will have to decide whether the writer is giving his or her own personal opinion, or presenting something as a fact. Section 1 may contain questions that will test your ability to identify what is presented as a fact and what is presented as an opinion.

Fact	Opinion
'There are 7 billion people in this world…'	'I believe there are more than 7 billion people in this world…'
'She is an Australian…'	'She sounded like an Australian…'
'Trump is the current President…'	'Trump is a horrible President…'
'Vegetables contain a lot of fibre…'	'Vegetables are good for you…'

Numerical and spatial reasoning
There are several questions that will test how well you can cope with numbers, and you should ideally be comfortable with simple mental calculations and being able to think logically.

Section 1B: An Overview

What will you be tested on?	No. of Questions	Duration
The ability to apply mathematical knowledge up to A Level	15 MCQs	80 Minutes (incl. Section 1A)

Section 1B of the exam involves short MCQ questions relating to Mathematics that are designed to see if you can quickly apply the principles that you have learnt in school in a time pressured exam. Assuming you split your time evenly between sections 1A + 1B (40 minutes each), you will have on average 160 seconds per question so it vital to work very quickly- some questions later tend to be harder so you should be doing the initial questions in under 90 seconds. I cannot emphasise enough that the limiting factor in this test is time not your ability. Practice is therefore crucial to learn the technique, skills and tricks to answer section 1 questions quickly. A quick summary of the syllabus is included below:

- **Number**- you should be confident performing a wide range of numerical calculations without the use of a calculator. As this is a MCQ exam, producing order of magnitude estimates will be very useful
- **Algebra & Functions**- you should be competent at basic algebra taught up to AS maths. You will already be at this standard but the key is to practice lots of questions so you use your algebra at the required speed The inequalities can be challenging to do under to time pressure so we recommend quickly drawing out the xy plane and identifying the region of interest. The factor/remainder theorem from A2 maths also appear in the syllabus so you may be tested on this.
- **Measure**- this is linked to "number" but we recommend that you become fully confident when dealing with scale factors. A question can often be simplified by working with this approach.
- **Statistics**- a very basic knowledge of GCSE statistics is all that is necessary. It is however important to know how to combine different statistics together and not get bogged down in long calculations
- **Probability**- a basic GCSE level of knowledge of probability but you will need to work through these questions quickly. We recommend that you practice drawing out 'tree diagrams' quickly to solve these problems
- **Coordinate Geometry** in the (x,y) plane- This is also content covered in AS Level maths and the challenge will be completing questions under time pressure. Practise converting equations to a standard form and then sketching them on the xy plane- this will often help you spot the solution. As this is a MCQ exam, you will not need to provide geometric proofs.
- **Trigonometry** - Basic trigonometry covering material tested in AS level maths. You are expected to know two basic trig formulae as well as the values of sine, cosine and tangent for the angles $0°, 30°, 45°, 60°, 90°$. As you will not have a calculator, it is crucial to memorise these values. This will also be very useful for interviews.
- **Exponential and Logarithms**- be confident at using the log formulae that you learnt in AS level maths. Using the formula will often simplify a question and with practice you will be able to determine whether to solve an equation in exponential form or logarithmic form
- **Differentiation**- when sitting your ENGAA exam, you will likely have covered advanced topics in differentiation including the product, chain and quotient rule. However, the exam itself only tests very basic differentiation taught in AS maths so try not to overcomplicate these questions.
- **Integration**- this once again contains the basic integration taught as part of AS level maths. It will be important to practice definite integration without a calculator as it is very easy to make a simple mistake.
- **Graphs of Functions**- this is a very important topic as it provides a lot of tricks to solve maths questions. We recommend you know the C3 transformations of graphs inside out and draw out sketches of common functions

Section 2: An Overview

What will you be tested on?	No. of Questions	Duration
Your ability to write an essay under timed conditions, your writing technique and your argumentative abilities	No Choice – Only one question	40 Minutes

Section 2 is usually what students are more comfortable with – after all, many GCSE and A Level subjects require you to write essays within timed conditions. It does not require you to have any particular legal knowledge – the questions can be very broad and cover a wide range of topics.

Here are some of the topics that might appear in Section two:

- Science
- Politics
- Religion
- Technology
- Ethics
- Morality
- Philosophy
- Education
- History
- Geopolitics

As you can see, this list is very broad and definitely non-exhaustive, and you do not get many choices to choose from (you have to write one essay out of three choices). Many students make the mistake of focusing too narrowly on one or two topics that they are comfortable with – this is a dangerous gamble and if you end up a topic you are unfamiliar with, this is likely to negatively impact your score.

You should ideally focus on at least four topics to prepare from the ECAA, and you can pick and choose which topics from the list above are the ones you would be more interested in. Here are some suggestions:

Economics Science
An essay that is related to science might relate to recent technological advancements and their implications, such as the rise of Bitcoin and the use of blockchain technology and artificial intelligence. This is interrelated to ethical and moral issues, hence you cannot merely just regurgitate what you know about artificial intelligence or blockchain technology. The examiners do not expect you to be an expert in an area of science – what they want to see is how you identify certain moral or ethical issues that might arise due to scientific advancements, and how do we resolve such conundrums as human beings.

Politics
Politics is undeniably always a hot topic and consequently a popular choice amongst students. The danger with writing a politics question is that some students get carried away and make their essay too one-sided or emotive – for example a student may chance upon an essay question related to Brexit and go on a long rant about why the referendum was a bad idea. You should always remember to answer the question and make sure your essay addresses the exact question asked – do not get carried away and end up writing something irrelevant just because you have strong feelings about a certain topic.

Religion
Religion is always a thorny issue and essays on religion provide strong students with a good opportunity to stand out and display their maturity in thought. Questions can range from asking about your opinion with regards to banning the wearing of a headdress to whether children should be exposed to religious practices at a young age. Questions related to religion will require a student to be sensitive and measured in their answers and it is easy to trip up on such questions if a student is not careful.

Education
Education is perhaps always a relatable topic to students, and students can draw from their own experience with the education system in order to form their opinion and write good essays on such topics. Questions can range from whether university places should be reduced, to whether we should be focusing on learning the sciences as opposed to the arts.

Section 2: Revision Guide

SCIENCE

Resource	What to read/do
1. Newspaper Articles	• The Guardian, The Times, The Economist, The Financial Times, The Telegraph, The New York Times, The Independent
2. A Levels/IB	• Look at the content of your science A Levels/IB if you are doing science subjects and critically analyse what are the potential moral/ethical implications • Use your A Levels/IB resources in order to seek out further readings – e.g. links to a scientific journal or blog commentary • Remember that for your LNAT essay you should not focus on the technical issues too much – think more about the ethical and moral issues
3. Online videos	• There are plenty of free resources online that provide interesting commentary on science and the moral and ethical conundrums that scientists face on a daily basis • E.g. Documentaries and specialist science channels on YouTube • National Geographic, Animal Planet etc. might also be good if you have access to them
4. Debates	• Having a discussion with your friends about topics related to science might also help you formulate some ideas • Attending debate sessions where the topic is related to science might also provide you with excellent arguments and counter-arguments • Some universities might also host information sessions for sixth form students – some might be relevant to ethical and moral issues in science
5. Museums	• Certain museums such as the Natural Science Museum might provide some interesting information that you might not have known about
6. Non-fiction books	• There are plenty of non-fiction books (non-technical ones) that might discuss moral and ethical issues about science in an easily digestible way

POLITICS

Resource	What to read/do
1. Newspaper Articles	• The Guardian, The Times, The Economist, The Financial Times, The Telegraph, The New York Times, The Independent
2. Television	• Parliamentary sessions • Prime Minister Questions • Political news
3. Online videos	• Documentaries • YouTube Channels
4. Lectures	• University introductory lectures • Sixth form information sessions
5. Debates	• Debates held in school • Joining a politics club
6. Podcasts	• Political podcasts • Listen to both sides to get a more rounded view (e.g. listening to both left and right wing podcasts)

RELIGION

Syllabus Point	What to read/do
1. Newspaper Articles	• The Guardian, The Times, The Economist, The Financial Times, The Telegraph, The New York Times, The Independent
2. Non-fiction books	• Read up about books that explain the origins and beliefs of different types of religion • E.g. Books that talk about the origins of Christianity, Islam or Buddhism, theology books etc.
3. Talking to religious leaders	• Talking to religious leaders may be a good way of understanding different religions more and being able to write an essay on religion with more maturity and nuance • Talking to people from different religious backgrounds may also be a good way of forming a more well-rounded opinion
4. Online videos	• Documentaries on religion • YouTube channels providing informative and educational videos on different religions – e.g. history, background
5. Lectures	• Information sessions • Relevant introductory lectures
6. Opinion articles	• Informative blogs and journals • Read both arguments and counter-arguments and come up with your own viewpoint

EDUCATION

Syllabus Point	What to read/do
1. Newspaper Articles	• The Guardian, The Times, The Economist, The Financial Times, The Telegraph, The New York Times, The Independent
2. A Levels/IB	• Draw inspiration from what you are studying in your A Levels or IB – do you feel like what you are studying is useful and relevant? E.g. Studying arts versus science • Compare the education you are receiving with your friends in different schools or different subjects
3. Educational exchange	• If you have an opportunity to go on an educational exchange, this might be a good opportunity to compare and contrast different educational systems • E.g. the approach to education in Germany versus the UK
4. University applications	• Have a read of how different universities promote themselves – do they claim to provide students with academic enlightenment, or better job prospects, or a good social life? • Why do different universities focus on different things?
5. Online videos	• Documentaries • YouTube Channels
6. Talk to your teachers	• Your teachers have been in the education industry for years and maybe decades – talk to them and ask them for their opinion • Talk to different teachers and compare their opinions regarding how we should approach education

Maths Revision Checklist

The material for the overviews of sections one and two have mainly been taken from the 2017 syllabus - this may change in the future. We recommend you consult the most up to date syllabus to see if there are any differences.

Syllabus Point	What to Know
1. Number	Understand and use BIDMAS Define; factor, multiple, common factor, highest common factor, least common multiple, prime number, prime factor decomposition, square, positive and negative square root, cube and cube root Use index laws to simplify multiplication and division of powers Interpret, order and calculate with numbers written in standard index form Convert between fractions, decimals and percentages Understand and use direct and indirect proportion; Apply the unitary method Use surds and π in exact calculations, simplify expressions that contain surds. Calculate upper and lower bounds to contextual problems Rounding to a given number of decimal places or significant figures
2. Algebra	Simplify rational expressions by cancelling or factorising and cancelling Set up quadratic equations and solve them by factorising Set up and use equations to solve problems involving direct and indirect proportion Use linear expressions to describe the nth term of a sequence Use Cartesian coordinates in all four quadrants Equation of a straight line, $y=mx+c$, parallel lines have the same gradient Graphically solve simultaneous equations Recognise and interpret graphs of simple cubic functions, the reciprocal function, trigonometric functions and the exponential function $y=kx$ for integer values of x and simple positive values of k Draw transformations of $y = f(x)$ [$y=af(x)$, $y=f(ax)$, $y=f(x)+a$, $y=f(x-a)$ only]
3. Geometry	Recall and use properties of angle at a point, on a straight line, perpendicular lines and opposite angles at a vertex, and the sums of the interior and exterior angles of polygons Understand congruence and similarity; Use Pythagoras' theorem in 2-D and 3-D Use the trigonometric ratios, between 0° and 180°, to solve problems in 2-D and 3-D Understand and construct geometrical proofs, including using circle theorems: a. **the angle subtended at the circumference in a semicircle is a right angle** b. **the tangent at any point on a circle is perpendicular to the radius at that point** Describe and transform 2-D shapes using single or combined rotations, reflections, translations, or enlargements, including the use of vector notation
4. Measures	Calculate perimeters and areas of shapes made from triangles, rectangles, and other shapes, find circumferences and areas of circles, including arcs and sectors Calculate the volumes and surface areas of prisms, pyramids, spheres, cylinders, cones and solids made from cubes and cuboids (formulae given for the sphere and cone) Use vectors, including the sum of two vectors, algebraically and graphically Discuss the inaccuracies of measurements; Understand and use three-figure bearings
5. Statistics	Identify possible sources of bias in experimental methodology Discrete vs. continuous data; Design and use two-way tables Interpret cumulative frequency tables and graphs, box plots and histograms Define mean, median, mode, modal class, range, and inter-quartile range Interpret scatter diagrams and recognise correlation, drawing and using lines of best fit Compare sets of data by using statistical measures
6. Probability	List all the outcomes for single and combined events Identify mutually exclusive outcomes; the sum of the probabilities of all these outcomes is 1 Construct and use Venn diagrams Know when to add or multiply two probabilities, and understand conditional probability Understand the use of tree diagrams to represent outcomes of combined events

How to use this Book

If you have done everything this book has described so far then you should be well equipped to meet the demands of the ECAA, and therefore **the mock papers in the rest of this book should ONLY be completed under exam conditions**.

This means:

- Absolute silence – no TV or music
- Absolute focus – no distractions such as eating your dinner
- Strict time constraints – no pausing half way through
- No checking the answers as you go
- Give yourself a maximum of three minutes between sections – keep the pressure up
- Complete the entire paper before marking
- Mark harshly

In practice this means setting aside two hours in an evening to find a quiet spot without interruptions and tackle the paper. Completing one mock paper every evening in the week running up to the exam would be an ideal target.

- Tackle the paper as you would in the exam.
- Return to mark your answers, but mark harshly if there's any ambiguity.
- Highlight any areas of concern.
- If warranted read up on the areas you felt you underperformed to reinforce your knowledge.
- If you inadvertently learnt anything new by muddling through a question, go and tell somebody about it to reinforce what you've discovered.

Finally relax… the ECAA is an exhausting exam, concentrating so hard continually for two hours will take its toll. So, being able to relax and switch off is essential to keep yourself sharp for exam day! Make sure you reward yourself after you finish marking your exam.

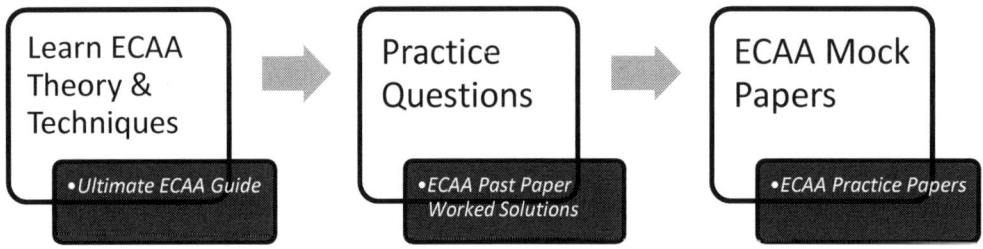

Scoring Tables

Use these to keep a record of your scores from past papers – you can then easily see which paper you should attempt next (always the one with the lowest score).

SECTION 1A	1st Attempt	2nd Attempt	3rd Attempt
Specimen			
2016			
2017			

SECTION 1B	1st Attempt	2nd Attempt	3rd Attempt
Specimen			
2016			
2017			

And the same again here but with our mocks instead.

SECTION 1A	1st Attempt	2nd Attempt	3rd Attempt
Mock A			
Mock B			

SECTION 1B	1st Attempt	2nd Attempt	3rd Attempt
Mock A			
Mock B			

Mock Paper A

Section 1A

Question 1

Competitors need to be able to run 200 metres in under 25 seconds to qualify for a tournament. James, Steven and Joe are attempting to qualify. Steven and Joe run faster than James. James' best time over 200 metres is 26.2 seconds. Which response is definitely true?

A. Only Joe qualifies
B. James does not qualify.
C. Joe and Steven both qualify
D. Joe qualifies
E. No one qualifies

Question 2

You spend £5.60 in total on a sandwich, a packet of crisps and a watermelon. The watermelon cost twice as much as the sandwich, and the sandwich cost twice the price of the crisps. How much did the watermelon cost?

A. £1.20
B. £2.60
C. £2.80
D. £3.20
E. £3.60

Question 3

Jane, Chloe and Sam are all going by train to a football match. Chloe gets the 2:15pm train. Sam's journey takes twice as long Jane's. Sam catches the 3:00pm train. Jane leaves 20 minutes after Chloe and arrives at 3:25pm. When will Sam arrive?

A. 3:50pm
B. 4:10pm
C. 4:15pm
D. 4:30pm
E. 4:40pm

Question 4

Michael has eleven sweets. He gives three sweets to Hannah. Hannah now has twice the number of sweets Michael has remaining. How many sweets did Hannah have before the transaction?

A. 11
B. 12
C. 13
D. 14
E. 15

Question 5

Alex gets a pay rise of 5% plus an extra £6 per week. The flat rate of income tax is decreased from 14% to 12% at the same time. Alex's current weekly take-home pay is £250 per week.
What will his new weekly take-home pay be, to the nearest whole pound?

A. £260
B. £267
C. £273
D. £279
E. £285

Question 6

You have four boxes, each containing two coloured cubes. Box A contains two white cubes, Box B contains two black cubes, and Boxes C and D both contain one white cube and one black cube. You pick a box at random and take out one cube. It is a white cube. You then draw another cube from the same box.
What is the probability that this cube is not white?

A. ½   B. ⅓   C. ⅔   D. ¼   E. ¾

Question 7

Anderson & Co. hire out heavy plant machinery at a cost of £500 per day. There is a surcharge for heavy usage, at a rate of £10 per minute of usage over 80 minutes. Concordia & Co. charge £600 per day for similar machinery, plus £5 for every minute of usage.
For what duration of usage are the costs the same for both companies?

A. 100 minutes   B. 130 minutes   C. 140 minutes   D. 170 minutes   E. 180 minutes

Question 8

Simon is discussing with Seth whether or not a candidate is suitable for a job. When pressed for a weakness at interview, the candidate told Simon that he is a slow eater. Simon argues that this will reduce the candidate's productivity, since he will be inclined to take longer lunch breaks.
Which statement **best** substantiates Simon's argument?

A. Slow eaters will take longer to eat lunch
B. Longer lunch breaks are a distraction
C. Eating more slowly will reduce the time available to work
D. Eating slowly is a weakness
E. People who like food are more likely to eat slowly

Question 9

Three pieces of music are on repeat in different rooms of a house. One piece of music is three minutes long, one is four minutes long and the final one is 100 seconds long. All pieces of music start playing at exactly the same time. How long is it until they are next all starting together?

A. 12 minutes   B. 15 minutes   C. 20 minutes   D. 60 minutes   E. 300 minutes

Question 10

A car leaves Salisbury at 8:22am and travels 180 miles to Lincoln, arriving at 12:07pm. Near Warwick, the driver stopped for a 14 minute break. What was its average speed, whilst travelling, in kilometres per hour? It should be assumed that the conversion from miles to kilometres is 1:1.6.

A. 51kph   B. 67kph   C. 77kph   D. 82kph   E. 386kph

Questions **11** and **12** refer to the following data:

Five respondents were asked to estimate the value of three bottles of wine, in pounds sterling.

Respondent	Wine 1	Wine 2	Wine 3
1	13	16	25
2	17	16	23
3	11	17	21
4	13	15	14
5	15	19	29
Actual retail value	8	25	23

Question 11

What is the mean error margin in the guessing of the value of wine 1?

A. £4.80   B. £5.60   C. £5.80   D. £6.20   E. £6.40

Question 12

Which respondent guessed most accurately on average?

A. Respondent 1   B. Respondent 2   C. Respondent 3   D. Respondent 4   E. Respondent 5

Questions **13** and **14** refer to the following data:

The population of Country A is 40% greater than the population of Country B.

The population of Country C is 30% less than the population of Country D (which is has a population 20% greater than Country B).

Question 13

Given that the population of Country A is 45 million, what is the population of country D?

A. 32.1 million   B. 35.8 million   C. 36.6 million   D. 38.6 million   E. 39.0 million

Question 14

The population of Country A is still 45 million. If Country B introduced a new health initiative costing $45 per capita, what would be the total cost?

A. $1.35 bn   B. $1.45 bn   C. $1.50 bn   D. $1.55 bn   E. $1.65 bn

Question 15

A car averages a speed of 30mph over a certain distance and then returns over the same distance at an average speed of 20mph. What is the average speed for the journey as a whole?

A. 22.5 mph   B. 24 mph   C. 25 mph   D. 26 mph

E. The distance travelled is required to calculate average speed

Question 16

"All sheep are ruminants and all marsupials are mammals. No sheep are marsupials." Which of the following must be true?

A. Some ruminants are marsupials.   B. All mammals are marsupials   C. All sheep are mammals

D. Some sheep are marsupials.   E. None of the above

Question 17

The price of toothpaste rises by 80%. This is later reduced by 50% due to competition. Zoe buys two tubes of toothpaste and gets the third free because of a loyalty card. How much did she have to pay per tube of toothpaste? Express your answer as a percentage of the original price.

A. 16.67%   B. 33%   C. 60%   D. 66.7%   E. 100%

Question 18

"You can remain fit throughout life if you exercise regularly. Simon does not exercise regularly, so he can never become fit." Which flawed argument has the same structure as this?

A. "You can speak a foreign language if you learn when young. Simon does not speak a foreign language, so he did not learn when young."

B. "You are never tired if you sleep for 8 hours a night. Simon is tired, therefore he doesn't sleep for 8 hours a night"

C. "You can be a good musician if you practice regularly. Simon does not practice regularly, so he can never be a good musician."

D. "You can be good at sport if you have a natural ability. Simon is good at hockey, therefore he has a natural ability."

E. "Eating five portions of fruit and vegetables daily reduces the risk of heart disease. Simon eats more than this, so he will not develop heart disease."

Question 19

"Reports of cybercrime are increasing year on year. Last year, police dealt with 250% more cybercrime then the year before. Common complaints relate to inappropriate or defamatory use of social media. To deal with this, many police forces are creating dedicated teams to deal with online offences. A pilot study showed that a dedicated cybercrime team solved cases of cybercrime 40% faster than regular detectives. Therefore the measure will act to suppress the rise in cybercrime."

Which statement best validates the above argument?

A. Solving crimes faster is necessary to keep pace with the increase in crime
B. Solving crimes faster leads to more convictions
C. Solving crimes faster increases police resources to tackle crime
D. Solving crimes faster saves money
E. Solving crimes faster reassures the public of action

Question 20

"Recently in Kansas, a number of farm animals have been found killed in the fields. The nature of the injuries is mysterious, but consistent with tales of alien activity. Local people talk of a number of UFO sightings, and claim extra terrestrial responsibility. Official investigations into these claims have dismissed them, offering rational explanations for the reported phenomena. However, these official investigations have failed to deal with the point that, even if the UFO sightings can be explained in rational terms, the injuries on the carcasses of the farm animals cannot be. Extra terrestrial beings must therefore be responsible for these attacks."

Which of the following best expresses the main conclusion of this argument?

A. Sightings of UFOs cannot be explained by rational means
B. Recent attacks must have been carried out by extraterrestrial beings
C. The injuries on the carcasses are not due to normal predators
D. UFO sightings are common in Kansas
E. Official investigations were a cover-up

Question 21

"To make a cake you must prepare the ingredients and then bake it in the oven. You purchase the required ingredients from the shop, however the oven is broken. Therefore you cannot make a cake."

Which of the following arguments has the same structure?

A. To get a good job, you must have a strong CV then impress the recruiter at interview. Your CV was not as good as other applicants, therefore you didn't get the job.

B. To get to Paris, you must either fly or take the Eurostar. There are flight delays due to dense fog, therefore you must take the Eurostar.

C. To borrow a library book, you must go to the library and show your library card. At the library, you realise you have forgotten your library card. Therefore you cannot borrow a book.

D. To clean a bedroom window, you need a ladder and a hosepipe. Since you don't have the right equipment, you cannot clean the window.

E. Bears eat both fruit and fish. The river is frozen, so the bear cannot eat fish.

Question 22

"Growing vegetables requires patience, skill and experience. Patience and skill without experience is common – but often such people give up prematurely as skill alone is insufficient to grow vegetables, and patience can quickly be exhausted."

Which of the following summarises the main argument?

A. Most people lack the skill needed to grow vegetables
B. Growing vegetables requires experience
C. The most important thing is to get experience
D. Most people grow vegetables for a short time but give up due to a lack of skill
E. Successful vegetable growers need to have several positive traits

Section 1B

Question 23

If the lines $y_1 = (n + 1)x + 10$ and $y_2 = (n + 3)x + 2$ are perpendicular then n must equal which of the following?

A. 2 B. -2 C. 3 D. -3 E. 0 F. 1

Question 24

The curve $y = x^2 + 3$ is reflected about the line $y = x$ and subsequently translated by the vector $\binom{4}{2}$. Which of the following is the x-intercept of the resulting curve?

A. -2 B. 11 C. 7 D. -11 E. 8 F. -8

Question 25

Given that $a^{3x}b^x c^{4x} = 2$, where a > 0, b > 0, and c > 0, then x =

A. $\frac{2}{3a+b+4c}$
B. $\frac{\log_{10} 2}{\log_{10}(a^3 bc^4)}$
C. $\frac{\log_{10} 2}{\log_{10}(a^3 bc^4)}$
D. $\log_{10} \frac{2}{(ab^2 c^3)}$
E. $\frac{\log_2 10}{\log_2(a^3 bc^4)}$
F. $\log_{10} \frac{2}{(a^3 bc^4)}$

Question 26

The sum of the roots of the equation $2^{2x} - 8 \times 2^x + 15 = 0$ is

A. 4
B. 16
C. $\log_{10}\left(\frac{15}{2}\right)$
D. $\frac{\log_{10} 15}{\log_{10} 2}$
E. 8
F. $\log_2\left(\frac{2}{3}\right)$

Question 27

For what values of the non-zero real number a does the equation $ax^2 + (a - 2)x = 2$ have real and distinct roots?

A. $a \neq -2$
B. $a > 2$
C. $a > -2$
D. No values of a.
E. $a \neq 0$
F. $a > 5$

Question 28

A bag only contains 2n blue balls and n red balls. All the balls are identical apart from colour. One ball is randomly selected and not replaced. A second ball is then randomly selected. What is the probability that at least one of the selected balls is red?

A. $\frac{4n}{3(3n-1)}$
B. $\frac{5n-1}{3(3n-1)}$
C. $\frac{5n-5}{3(3n-1)}$
D. $\frac{4n-2}{3(3n-1)}$
E. $\frac{n-5}{9(n-1)}$
F. $\frac{4n-1}{3(3n-1)}$

Question 29

Which of the following equations is a correct simplification of the equation $\frac{x^2-16}{x^2-4x}$?

A. $1-\frac{4}{x}$
B. $\frac{x+4}{x}$
C. $\frac{x-4}{x}$
D. $\frac{4}{x}$
E. $\frac{x(x-4)}{x}$
F. $\frac{x+4}{4x}$

Question 30

What is the equation of the quadratic function that passes through the x-coordinates of the stationary points of $y = x^2 e^x$?

A. Function does not exist.
B. $x^2 + 2x$
C. x^2
D. $x^2 - 2x$
E. $x^2 + 4x$
F. $2x^2 - 1$

Question 31

Given a curve with the equation $y = 8 - 4x - 2x^2$ and a line $y = k(x + 4)$, find the values of k for which the line and the curve are tangent to each other.

A. $-4 < k \leq 4$
B. $k = 4, k = 20$
C. $4 < k < 20$
D. $k = -4, k = 4$

Question 32

Given the two equations $y_1 = (1 - x)^6$ and $y_2 = (1 + 2x)^6$, find the ratio of the coefficients of the 2nd term in the expansion of y_1 and the 3rd term in the expansion of y_2 (The y_1 coefficient should be the numerator, and the y_2 coefficient should be the denominator).

A. $\frac{-1}{10}$
B. $\frac{1}{9}$
C. $\frac{1}{15}$
D. $-\frac{1}{7}$.

Question 33

What is the sum of the integers from 1 to 300?

A. 9,000 B. 44,850 C. 45,150 D. 45,450 E. 54,450 F. 90,000

Question 34

If $sin2\theta = \frac{2}{5}$, then what is $\frac{1}{sin\theta cos\theta}$?

A. $\frac{1}{5}$ B. $\frac{5}{4}$ C. $\frac{5}{2}$ D. 5 E. $\frac{3}{2}$ F. 1

Question 35

If $[n]$ represents the greatest integer less than or equal to n, then which of the following is the solution to $-11 + 4[n] = 5$?

A. n = 4 B. 4<n<5 C. -2≤n≤-1 D. 4≤n<5 E. 4<n≤5 F. n<5

Question 36

The operation Ø is defined for all real numbers a and b(b ≠ 0) as $a\emptyset b = \frac{a/2}{b}$.

If $10\emptyset n = n\emptyset \left(\frac{1}{10}\right)$, which of the following is a solution for n?

A. 1 B. $5\sqrt{2}$ C. -10
D. 10 E. $-5\sqrt{2}$ F. -1

Question 37

If -1 is a zero of the function f(x) = $2x^3 + 3x^2 - 20x - 21$, then what are the other zeroes?

A. 1 and 3 B. -3 and 3 C. $\frac{-7}{2}$ and 1 and 3
D. $\frac{-7}{2}$ and 3 E. $-1 \wedge 3$ F. 1 and 7

Section 2

Read the extract taken from Neal Reaich's *What price honesty?* (2014, Bized) and then answer the question below in the space provided in this booklet.

Your answer will be assessed taking into account your ability to construct a reasoned, insightful and logically consistent argument with clarity and precision.

QUESTION
To what extent can prices be considered a true reflection of value, and to what extent should governments regulate price-setting behaviours such as reference pricing? Discuss with reference to the passage above.

What price honesty?

Pricing seems to be mentioned quite a lot in recent news headlines. The latest is the investigation by the Office of Fair Trading (OFT) into just how genuine the advertised price cuts in some large furniture stores really are. The OFT use the term reference pricing. They've found cases in shops under investigation where not a single product had, in reality, been sold at the, supposedly original, higher price. The argument goes that since 95% of sales were at the lower or 'now' price then the stated original prices cannot be really genuine.

Are we really so stupid as to be misled by reference pricing? When my wife buys some clothes in a sale and I ask how much it cost, she always responds by saying how much money she has saved. Well, unless she had the definite intention of buying it in the first place, she hasn't saved anything: in fact she has spent it. The question is whether or not the reference pricing made a difference to her decision to buy the product. It must do, otherwise why is it so commonly used?

Heuristics suggests that people are not as rational as the standard economic model implies. Instead they use rules of the thumb, educated guesses or short cuts in decision-making. Anchoring refers to people making decisions based upon something they know to start with. For example the anchor price of a jumper was £40 and there is 20 per cent discount offer making the new price £32 a saving of £8. If I decide to buy I might be thinking of the £8 saved because I didn't have to pay the full price, when I should be thinking logically about the actual price I am paying. In this case there is too much focus on the anchor price, which is at a level that I wouldn't have purchased the item anyway.

In some cases reference pricing is part of a strategy of price discrimination over time. Clothes shops attempt to capture consumer surplus by charging a high initial price for a few weeks for the 'must have and will pay' customers. The shops then give a discount which increases over time. New potential customers must now weigh up whether to buy now and get a 25 per cent discount or wait longer for a 5per cent saving and risk losing the deal because they have run out of their size.

Another tactic is to use time-limited offers where notice is given that the offer ends soon: buy now or miss out. Double glazing sales were notorious for using this tactic, encouraging customers to 'sign up today to will get an extra 20 per cent off'.

Supermarkets are always giving volume offers, such as three for the price of two or get the second purchase for half the price. The supermarkets know we will pay more for the first item than we would for a second or third one. This tactic must lead to food waste as we are tempted to buy some products that we cannot possibly use before the sell-by date.
The OFT also look at baiting sales where only a very limited number of products are available at the most discounted price. This sounds a little like discounts given for advanced booking train tickets.

The OFT always makes mention of portioned 'drip' pricing where price increments drip through the buying process. It's those little add-ons that all add-up. Booking certain airline tickets comes to mind here.

Restaurants often make an offer of a free second main course meal after buying one main course at full price. Now add on the fact that we may have full priced deserts, order drinks at a high mark up and then add the tip. Get your calculator out and the deal doesn't seem so good.

All these examples of price framing seek to alter a consumer's perception of the value of the offer. But there is more: have you purchased a printer at a ridiculously low price only to be stung on purchasing printer ink cartridges? Some businesses operate at a loss or low profit margin on certain items, to entice customers to part with more money on higher-profit goods. They bundle items together.

Some advice then:
- Use a calculator with you when shopping
- When buying clothes divide the total by 50 to give you the average cost of wearing something once a week for a year: a shirt costing £25 will work out at 50 pence for every day worn;
- Avoid impulse buying on larger-ticket items;
- Only purchase multi-buys with long sell buy dates;
- Ignore the original price: it's only the current price you need to know about;
- Avoid time limited offers.

END OF PAPER

Mock Paper B

Section 1A

Question 1

Joseph has a bag of building blocks of various shapes and colours. Some of the cubic ones are black. Some of the black ones are pyramid shaped. All blue ones are cylindrical. There is a green one of each shape. There are some pink shapes. Which of the following is definitely **NOT** true?

A. Joseph has pink cylindrical blocks
B. Joseph doesn't have pink cylindrical blocks
C. Joseph has blue cubic blocks
D. Joseph has a green pyramid
E. Joseph doesn't have a black sphere

Question 2

Sam notes that the time on a normal analogue clock is 1540hrs. What is the smaller angle between the hands on the clock?

A. 110°   B. 120°   C. 130°   D. 140°   E. 150°

Question 3

A fair 6-faced die has 2 sides painted red. The die is rolled 3 times. What is the probability that at least one red side has been rolled?

A. $8/27$ B. $19/27$ C. $21/27$ D. $24/27$ E. 1

Question 4

In a particular furniture warehouse, all chairs have four legs. No tables have five legs, nor do any have three. Beds have not less than four legs, but one bed has eight as they must have a multiple of four legs. Sofas have four or six legs. Wardrobes have an even number of legs, and sideboards have an odd number. No other furniture has legs. Brian picks a piece of furniture out, and it has six legs.

What can be deduced about this piece of furniture?

A. It is a table
B. It could be either a wardrobe or a sideboard
C. It must be either a table or a sofa
D. It must be either a table, a sofa or a wardrobe
E. It could be either a bed, a table or a sofa.

Question 5

Two friends live 42 miles away from each other. They walk at 3mph towards each other. One of them has a pet pigeon which starts to fly at 18mph as soon as the friends set off. The pigeon flies back and forth between the two friends until the friends meet. How many miles does the pigeon travel in total?

A. 63   B. 84   C. 114   D. 126   E. 252

Question 6

"Fruit juice contains fibre, vitamins and minerals and can be part of a healthy diet. However, it has been suggested that the high sugar content and acidity negates these benefits by leading to increased rates of dental cavities and hyperactivity in children. If left unchecked, a combination of poor dental hygiene and inappropriate diet can lead to disastrous consequences, including serious infections. On the other hand, many juices contain essential vitamins such as vitamin C which helps the immune system fight infections."

What is the main message from this passage?

A. Children should not drink fruit juice
B. Fruit juice is harmful to health
C. Fruit juice is good for health
D. On balance, we should drink more fruit juice
E. The overall benefits of fruit juice are unclear

Question 7

A complete stationery set includes a pen, a pencil, a geometry set and a pad of paper. Pens cost £1.50, pencils cost 50p, geometry sets cost £3 and paper pads cost £1. Sam, Dave and George each want complete sets, but Mr Browett persuades them to share. Sam and Dave agree to share a paper pad and a geometry set. George must have his own pen, but agrees that he and Sam can share a pencil.

What is the total amount spent?

A. £12.00   B. £13.50   C. £16.50   D. £17.50   E. £18.00

Question 8

"If the government financially supports the arts, a proportion of each person's taxes will be used to finance museums, galleries and theatres. But some taxpayers have no interest in the arts and never go to theatres or museums. Many of those who enjoy the arts are able to afford to pay for them. Since no one should be forced to subsidise services which they themselves do not use, taxpayers' money should not be used to support the arts."

Which counter-argument provides the strongest rebuke of this principle?

A. If public funding for the arts is withdrawn, only those who are genuinely interested would pay to visit museums

B. The rail network is publically subsidised, although some people do not use trains

C. If people only pay for services they use, then those who can afford private health insurance would not pay towards the NHS

D. Funding museums allows greater preservation of our heritage

E. If something requires subsidy, then people must not genuinely want it

Question 9

The figure to the right shows 5 squares made from 12 matches. Which 2 matches need to be moved to make 7 squares?

A. 1 and 2   B. 1 and 3   C. 1 and 4

D. 3 and 5   E. Not possible

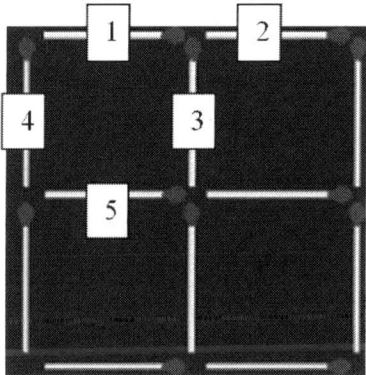

Question 10

A cube has six sides of different colours. The red side is opposite to black. The blue side is adjacent to white. The brown side is adjacent to blue. The final side is yellow. Which colour is opposite brown?

A. Red   B. Black   C. Blue   D. White   E. Yellow

Question 11

The UK imports 36,000,000kg of cocoa beans each year. Each g costs the UK 0.3p, from which the supplier takes 20% commission. Of what is left, the local government takes 60% and the distribution company gets 30%. How much are the cocoa farmers left with per year?

A. £3.68m   B. £6.82m   C. £8.64m   D. £10.8m   E. £11.4m

Questions **12** and **13** refer to the following passage:

- In the year ending June 2013 there were 1,730 fatalities in reported personal injury accidents, a 3 per cent drop from 1,785 in the year ending June 2012. The number of killed or seriously injured (KSI) casualties fell by 5 per cent, to 23,530, and the total number of casualties fell by 7 per cent to 188,540.
- A total of 8,560 car users were reported killed or seriously injured in the year ending June 2013, a fall of 6 per cent from the previous 12-month period.
- KSI casualties for the vulnerable road user groups – pedestrians, pedal cyclists and motorcyclists – showed overall decreases of 7, 1 and 6 per cent respectively compared with the year ending June 2012.
- The casualty rate per billion vehicle miles decreased for all casualty severities in the year ending June 2013, with falls of 3 per cent for fatalities, 6 per cent for serious injuries and 7 per cent for all casualties. This is the first publication in which the Department has included quarterly casualty rates.
- There were also significant decreases in the number of child casualties (aged 0-15) which fell from 18,166 in the year ending June 2012, to 15,920 in the year ending June 2013, a fall of 12 per cent. The number of child KSIs also fell in the same period by 11 per cent to 2,080. The number of child pedestrian casualties who were killed or seriously injured fell by 8 per cent to 1,440 in the year ending June 2013.
- There were drops in the number of accidents on all road types in the year ending June 2013 relative to the year ending June 2012. The number of fatal or serious accidents fell by 7 per cent on major roads (motorways and A roads) and 4 per cent on minor roads. On roads with speed limits over 40 mph (non-built up) fatal and serious accidents fell by 6 per cent and on roads with speeds limits up to an including 40 mph (built-up) they fell by 5 per cent.
- There were 185,540 casualties from 139,350 accidents in the year ending June 2013 which represents a 6 per cent fall for accidents and a 7 per cent fall for casualties compared with the year ending June 2012.

Question 12

Regarding the passage, which of these statements can be known to be true?

A. Child casualties are on the rise
B. Annual road deaths in the UK are falling
C. Vulnerable road users are more likely to be injured per vehicle mile than drivers
D. From June 2012 to June 2013, there were 188,540 serious injuries
E. Motorways are safer than built-up roads

Question 13

"The government is always under pressure to reduce road casualties. For this reason, anti-drink-drive campaigns costing millions of pounds are commonly produced, particularly around Christmas time. To address a one-year increase in drink driving related deaths, a new campaign was introduced. Subsequently, drink-driving casualties fell. The government therefore concluded that the £8m campaign had been a success."

Which of the following most undermines this argument?

A. Fewer people drink-drive these days than 10 years ago
B. Correlation does not imply causation: there is no plausible mechanism for the campaign to provide benefit.
C. The effect is too rapid for this campaign to have changed the public's attitude
D. Regression to the mean explains this phenomenon: values which were abnormally high one year are likely to settle down the next
E. When spending so much money, benefits are certain. The true test is in running a smaller campaign.

Question 14

"Some people with a sore throat and a chest infection have the 'flu."

Which of the following statements is supported?

A. Some people have a chest infection, but do not have the 'flu
B. Some people with a sore throat and a chest infection do not have the 'flu
C. Kate has the 'flu. Therefore she has a sore throat
D. The 'flu is defined as a sore throat and chest infection together
E. None of the above

Question 15

Catherine has 6 pairs of red socks, 6 pairs of blue socks and 6 pairs of grey socks in her drawer. Unfortunately, they are not paired together. The light in her room is broken so she cannot see what colour the socks are. She decides to keep taking socks from the drawer until she has a matching pair.

What is the minimum number of socks she needs to take from the drawer to guarantee at least one matching pair can be made?

A. 2   B. 3   C. 4   D. 5   E. 6

Question 16

Luca and Giovanni are waiters. One month, Luca worked 100 hours at normal pay and 20 hours at overtime pay. Giovanni worked 80 hours at normal pay and 60 hours at overtime pay. Neither received any tips. Luca earned €2000; Giovanni earned €2700. What is the overtime rate of pay?

A. €10 per hour   B. €15 per hour   C. €20 per hour   D. €25 per hour   E. €30 per hour

Question 17

"Train A leaves Plymouth at 10:00 and travels at 90mph. Train B leaves Manchester at 10:30 and travels at 70mph. The distance between the two cities is 405 miles. Due to a mistake, both trains are travelling on the same track."
Calculate the distance from Plymouth at which the trains will collide.

A. 158 miles   B. 203 miles   C. 228 miles   D. 248 miles   E. 263 miles

Question 18

"100 pieces of rabbit food will feed one pregnant rabbit and two normal rabbits for a day. 175 pieces of food will feed two pregnant and three normal rabbits for a day. There is no excess food."
Which statement is **NOT** true?

A. A normal rabbit can be fed for longer than a day with 30 pieces of food.
B. 70 pieces of food are sufficient to feed a pregnant rabbit for a day.
C. A pregnant rabbit needs twice as many pieces per day as a normal rabbit.
D. Two pregnant and four normal rabbits will need 200 pieces of food for a day.
E. Three pregnant and ten normal rabbits will need 450 pieces of food for a day.

Question 19

"Studies of the brains of London taxi drivers show that training for "the knowledge", a difficult exam requiring knowledge of 20,000 London streets, enlarged a part of the brain believed to be important for spatial and organisational memory. This shows the brain can adapt to training and increase its abilities. Therefore if I wanted to improve my ability to remember names, I should also train my brain with repetitive tasks."
Which of the following **best** represents the flaw in this argument?

A. Enlarging of the brain does not necessarily mean it has improved
B. It might not be true to assume name memory and spatial memory use the same part of the brain
C. The brain enlargement would likely have happened anyway even without training
D. We do not know how London taxi drivers prepare for "the knowledge"
E. Practice does not necessarily improve performance on memory tasks

Question 20

"Michael bought a painting at an auction for £60. After 6 months, he realised the value of the painting had increased, so he sold it for £90. Realising a mistake, he wanted to buy the painting back, which he was able to do for £110. A year later, he then re-sold the painting for £130."
What is the total profit on the painting?

A. £20   B. £30   C. £40   D. £50   E. £60

Question 21

"Insect pests such as aphids and weevils can be a problem for farmers, as they feed on crops, causing destruction. Thus, many farmers spray their crops with pesticides to kill these insects, increasing their crop yield. However, there are also predatory insects such as wasps and beetles that naturally prey on these pests – which are also killed by pesticides. Therefore, it would be better to let these natural predators control the pests, rather than by spraying needless chemicals." Which of the following best describes the flaw in this logic?

A. Many pesticides are expensive, so should not be used unless necessary
B. It fails to consider other problems the pesticides may cause
C. It does not explain why weevils are a problem
D. It fails to assess the effectiveness of natural predators compared to pesticides
E. It does not consider the benefits of using fewer pesticides

Question 22

A parliament contains 400 members. Last election, there was a majority of 43% of the popular vote to the liberal party. However, as a first-past-the-post system of constituencies was in effect, they gained 298 seats in parliament. How many excess members did they have, relative to a straight proportional representation system?

A. 72   B. 98   C. 112   D. 126   E. 148

END OF SECTION

Section 1B

Question 23

Consider the infinite series, $x - \left(\frac{1}{2}\right)x^2 + \left(\frac{1}{4}\right)x^3 - \left(\frac{1}{8}\right)x^4 \ldots$ Given that we know that the fifth term of the series is $\left(\frac{1}{32}\right)$, what is summation of the series given that the series converges as it heads toward infinity?

A $\dfrac{16^{\frac{1}{5}}}{2+\frac{(16^{\frac{1}{5}})}{2}}$

B $\dfrac{1}{1-(32)^{\frac{1}{4}}}$

C $\dfrac{8^{\frac{1}{5}}}{1+8^{\frac{1}{5}}}$

D $\dfrac{2}{2-(16)^{\frac{1}{4}}}$

E $\dfrac{-2}{2+(16)^{\frac{1}{4}}}$

F $\dfrac{1}{64-8^{\frac{1}{5}}}$

Question 24

If $\log_2 3 \cdot \log_3 4 \cdot \log_4 5 \ldots \log_n(n+1) \leq 10$, what is the largest value of n that satisfies this equation?

A. 1022   B. 824   C. 842   D. 1023   E. 1020   F. 890

Question 25

a,b,c is a geometric progression where a,b,c are real numbers. If $a+b+c=26$ and $a^2+b^2+c^2=364$, find b.

A. $\dfrac{1}{3\sqrt{25}}$ B. 6   C. $2\sqrt{6}$ D. 9   E. 4   F. $2\sqrt{3}$

Question 26

Given that a>0, find the value of a for which the minimal value of the function $f(x) = (a^2+1)x^2 - 2ax + 10$ in the interval $x \in [0; 12]$ is $\dfrac{451}{50}$.

A. 7   B. 12   C. 5   D. $\dfrac{50}{125}$ E. 8   F. 10

Question 27

If the probability that it will rain tomorrow is $\dfrac{2}{3}$ and the probability that it will rain and snow the following day is $\dfrac{1}{5}$, given that the probability of rain and snow occurring on any given day are independent from one another, what is the probability that it will snow the day after tomorrow?

A. $\dfrac{10}{3}$ B. $\dfrac{3}{10}$ C. $\dfrac{2}{15}$ D. $\dfrac{15}{2}$ E. $\dfrac{4}{9}$ F. $\dfrac{1}{5}$

Question 28
If $cos2\theta = \frac{3}{4}$, then $\frac{1}{cos^2\theta - sin^2\theta} =$

A. $\frac{4}{3}$ B. 4   C. -1 D. $\frac{3}{4}$ E. 2   F. 1

Question 29
Describe the geometrical transformation that maps the graph of $y = 0.2^x$ onto the graph of $y = 5^x$.

A. Reflection in the x-axis   B. Reflection in the y-axis   C. Multiplication by a scale factor of 25

D. Addition of the constant term 4.8   E. Multiplication by scale factor of 5   F. Multiplication by scale factor $\frac{1}{25}$

Question 30
Find the solution to the equation $log_4(2x + 3) + log_4(2x + 15) - 1 = log_4(14x + 5)$

A. There is no solution   B. $\frac{2}{5}$ C. $\frac{5}{2}$ D. -1   E. 1   F. 0

Question 31
The normal to the curve $y = e^{2x-5}$ at the point $P(2, e^{-1})$ intersects the x-axis at the point A and the y-axis at the point B. Which of the following is an appropriate formula for the area of the triangle that is formed in terms of e, m, and n, where m and n are integers?

A. $\frac{(e^2+1)^m}{e^n}$

B. $\frac{(e^3+1)^{\frac{1}{n}}}{m}$

C. $\frac{e^n}{(e^2+1)^m}$

D. $\frac{m^{\frac{1}{n}}}{e^3+1}$

E. $\frac{e^{2m}}{e^n+1}$

F. $\frac{(e^2-1)^m}{e^{2n}}$

Question 32
Given that $secx - tanx = -5$, find the value of cos x.

A. -0.2   B. 0.2   C. $\frac{-13}{5}$ D. $\frac{-5}{13}$ E. 0.5   F. -0.5

Question 33

Consider the line with equation $y = 2x + k$ where k is a constant, and the curve $y = x^2 + (3k - 4)x + 13$. Given that the line and the curve do not intersect, what are the possible values of k?

A. $\frac{-1}{3} < k < 3$ B. $\frac{-4}{9} < k < 4$ C. $\frac{1}{2} < k < \frac{5}{3}$

D. $\frac{3}{2} < k \leq \frac{8}{3}$ E. $\frac{1}{3} < k < 3$ F. $-3 < k < \frac{1}{3}$

Question 34

A circle with centre C(5,-3) passes through A(-2,1), and the point T lies on the tangent to the circle such that AT = 4. What is the length of the line CT?

A. 9   B. 18   C. $\sqrt{95}$ D. $8\sqrt{2}$ E. $\sqrt{69}$ F. 8

Question 35

Evaluate: $(6 \sin x)(3 \sin x) - (9 \cos x)(-2 \cos x)$

A. 0   B. 0.5   C. 1   D. -1   E. 18   F. -18

Question 36

In the figure to the right, all triangles are equilateral. What is the shaded area of the figure in terms of r?

A. $5r^2(2\sqrt{6} - 3\pi)$
B. $5r^2(5\sqrt{2} - 6\pi)$
C. $5r^2(3\sqrt{3} - \pi)$
D. $5r^2(4\sqrt{3} - 2\pi)$
E. $5r(2\sqrt{6} - 3\pi)$
F. $5r^2(5\sqrt{2} + 6\pi)$

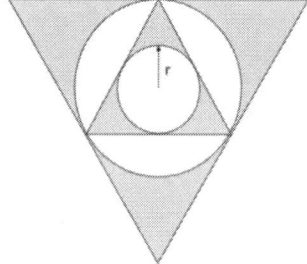

Question 37

Suppose I use a binomial expansion to determine the value of $(3.12)^5$. What is the minimum number of terms that I must obtain in the expansion of $(3.12)^5$ in order to receive a result accurate to 1 decimal place?

A. 4   B. 5   C. 6   D. 7   E. 9   F. 8

Section 2

Read the extract taken from Simon Tait's *Maria Miller: thank you and goodbye* (2014, *The Stage*- edited) and then answer the question below in the space provided in this booklet.

Your answer will be assessed taking into account your ability to construct a reasoned, insightful and logically consistent argument with clarity and precision.

To what extent does art and culture merit public funding?

Maria Miller: thank you and goodbye

She "has done an effective job in making the case for the value of public funding" Bazalgette says in response to the news that the arts are to get a ring-fenced 5% cut when ACE had been told to model for 10% and 15% scenarios for 2015/16.

But it is rather damning with faint praise. He goes on to say: "It is hugely encouraging that the Chancellor and the Treasury have listened to the argument that the arts and culture makes such a valuable contribution to our quality of life and the economy".

The argument made by the arts and culture, note, not the secretary of state.

DCMS (Department for Culture, Media and Sport) as a whole has got a less generous cut of 8%, and the deal for the arts appears to have been negotiated separately by the likes of Bazalgette and national museum directors like Nicholas Serota who, three weeks ago, went to George Osborne directly with economic arguments for a more lenient treatment of the sector. It was at this point that Osborne and the Treasury finally "got it" and realised how damaging a bigger arts cut would be to the economy for negligible saving.

It means that Mrs Miller cannot simply pass on to the arts the 8% cut as she and her predecessor, Jeremy Hunt, have done in the past because there is no fat in the DCMS, having been cut to the bone already, to absorb a new reduction itself. As it is, she will have to find the saving from elsewhere in her budget.

Nevertheless, she has hung out and got a better settlement than most other government departments who are suffering at least 10% reductions as the government tries to find more savings, but it seems the knives are out not for culture or the arts but for Miller herself.

The knives appear to be out for her in government for a number of reasons, including not dealing decisively with Leveson. The culture secretary has also been the subject of unprecedented vilification in the Tory press, with the Daily Mail's drama critic Quentin Letts declaring a couple of weeks ago that "Culture is the department where a country can assert its character. If only its Secretary of State had one". In May, she made her first speech on the arts, calling for the economic argument to be made, Letts conceded, but "Where was the question of morality in Mrs Miller's approach to the arts? Where was the vision that the arts can civilise us? Where was an idea of the arts as the most meritocratic of gifts, a route which can offer talented and aspiring youngsters a route to self-fulfilment…? There is not even much impression she is an arts lover. It was a speech that could have been given by any one of her departmental officials".

Her desperate attempts to grab a positive headline culminated last week in a damp squib of an announcement about the First World War centenary commemoration, in which nothing new was announced (except that 600-odd streets in England were to be renamed after VC winners from the Great War), and the major news about the cultural element cannot be revealed before August. On Friday, The Times's normally gentle columnist Richard Morrison wrote that "Some (culture secretaries) have been bores; some bluffers. But not one has depressed me as Maria Miller does.

As for the arts, the triumph is substantial and this might be a seachange in the way governments see the sector. The Arts Council, as fuel for the Bonfire of the Quangos, has taken an enormous battering since 2010 and the sector has correctly acknowledged that there is no case for "special treatment" while cuts amounting to 33% have been meted out, and of 50% to ACE itself. But now culture has established the principle that it is a special case after all, and with sense and imagination much of the effect of the new 5% cut might be ameliorated through the National Lottery.

The question now is whether that principle will be accepted by the other great subsidisers of the arts, the local authorities in whose hands the futures of dozens of theatres lie and whose extreme economic pain is even greater than Osborne's.

END OF PAPER

ANSWERS

Answer Key

Paper A		Paper B	
1	B	1	C
2	D	2	C
3	E	3	B
4	C	4	D
5	C	5	D
6	C	6	E
7	E	7	B
8	C	8	C
9	D	9	C
10	D	10	D
11	C	11	C
12	C	12	B
13	D	13	D
14	B	14	E
15	B	15	C
16	E	16	D
17	C	17	D
18	C	18	E
19	C	19	B
20	B	20	D
21	C	21	D
22	E	22	D
23	B	23	A
24	B	24	D
25	C	25	B
26	D	26	A
27	A	27	B
28	B	28	A
29	B	29	B
30	B	30	C
31	B	31	A
32	A	32	D
33	C	33	B
34	D	34	A
35	D	35	E
36	A	36	C
37	D	37	A

Mock Paper A Answers

Section 1A

Question 1: B
James runs 26.2 seconds, which is outside the qualifying time, therefore he does not qualify

Question 2: D
5.6/7 gives the unit price of 80p – this equals a packet of crisps. Multiplying this by 2 gives the sandwich and by 4 gives the watermelon price of £3.20

Question 3: E
Jane leaves at 2:35pm and arrives at 3:25pm, taking 50 minutes. Sam's journey takes twice as long, so leaving at 3:00pm it takes 100 minutes, giving an arrival time of 4:40pm

Question 4: C
After the donation, Sam has eight sweets. Therefore Hannah had 16 sweets after the transaction and hence 13 sweets before

Question 5: C
Find original pay: £250/0.86 = 290 basic original pay. Add the rise: (290 x 1.05) + 6 = £311 new basic pay. Subtract the income tax at 12% = 311 x 0.88 = £273 new pay rate

Question 6: C
Given the first cube is a white cube, you are drawing from one of three boxes, boxes A, C or D. Boxes C and D will have just had their only white cube removed, whereas box A will have one white cube remaining. Therefore the probability of drawing a second white cube is $1/3$, thus the probability of non-white (i.e. black) is $2/3$.

Question 7: E
This is a simultaneous equations question. $500 + 10(x - 80) = 600 + 5x$; true when $x \geq 80$.
$500 + 10x - 800 = 600 + 5x$
» $5x = 900$
» $x = 180$

Question 8: C
If eating more slowly caused a reduction in the time available to work, the candidate might be less productive.

Question 9: D
This is a LCM question. We need to find the lowest common multiple of the song lengths. The LCM of 100, 180 and 240 is 3,600 seconds – equal to 60 minutes. For ease of arithmetic, you may choose to work reduce all numbers by a factor of 10.

Question 10: D
The journey is 3 hours and 45 mins, minus a 14 minute break gives 3hrs 31 mins travel time, or 211 minutes. Therefore the average speed is 51mph, or 82kmh by using the stated conversion factor.

Question 11: C
The mean guess is £13.80, which is £5.80 too high.

Question 12: C
The overall error for respondent 3 is £13, which is the least

Question 13: D
Scale back and forth from known quantities. Country B has 32.1m so Country D has 38.6m people.

Question 14: B
Country B has 32.1m people. Therefore $45 x 32.1m = $1.45bn

Question 15: B
The average speed is 24mph, independent of distance travelled as it cancels. Imagine this covers a set distance of say 30 miles. It will take 1 hour on the way and 1.5 hours on the way back. 60/2.5 = 24. This is true of all distances, the ratio is the same.

Question 16: E
None of the above can be reliably deduced from the passage alone

Question 17: C
Imagine the toothpaste costs 100p originally, and follow the price through. It rises by 80% to 180p, then is reduced by 50% to 90p. Three tubes are purchased for the price of 2 (i.e. 180p), therefore the cost per unit is 180/3 = 60p. 60p = 60% x 100, the original price

Question 18: C
Argument C has the same form, asserting that since something is not happening, the result of the action will never be true.

Question 19: C
Statement C is the only one making reference to the potential outcome of solving crimes faster, thereby providing a plausible mechanism for a reduction in cybercrime rates

Question 20: B
The passage suggests that the attacks were carried out by extra terrestrial beings. Though the supposed UFO sightings have rational explanations, the writer feels this is insufficient to dismiss his idea.

Question 21: C
The initial argument suggests that two things must be present for an action to happen. If only one is absent, the action cannot happen. Argument C has the same form, the others do not.

Question 22: E
Growing vegetables needs several positive traits. The passage does not tell us which is the most important or most commonly lacked skill, only that more than one skill is required for success.

END OF SECTION

Section 1B

Question 23: B
We know that the product of slopes of perpendicular lines equals -1.
Therefore:
$(n + 1)(n + 3) = -1$.
$n^2 + 4n + 3 = -1$
$n^2 + 4n + 4 = 0$
Factorising gives (n+2)(n+2), therefore n = -2 for the lines to be perpendicular.

Question 24: B
Algebraically, we can find the result of reflecting the curve $y = x^2 + 3$ across the line y=x by replacing y with x in the equation, and solving for the value of y in order to find the relevant equation, which is:

$x = f(y) = \sqrt{y - 3}$
Replacing y with x gives:
$y = \sqrt{x - 3}$
Translating the resulting equation by $\binom{4}{2}$ corresponds to introducing (-4) to the x term and (+2) to the y:
$y + 2 = \sqrt{x - 4 - 3}$
$y = \sqrt{x - 7} + 2$
The x-intercept is found by setting f(x) = 0.
$\sqrt{x - 7} + 2 = 0$
$\sqrt{x - 7} = -2$
$x - 7 = 4$
$x = 11$

Question 25: C
Take logs of each side and separate out the LHS:
$3x \log_{10} a + x \log_{10} b + 4x \log_{10} c = \log_{10} 2$
$x (3 \log_{10} a + \log_{10} b + 4 \log_{10} c) = \log_{10} 2$
$x \log_{10}(a^3 b c^4) = \log_{10} 2$
$x = \frac{\log_{10} 2}{\log_{10}(a^3 b c^4)}$

Question 26: D
Let $y = 2^x$. Then, $y^2 - 8y + 15 = 0$.
Solving this either using the quadratic equation or otherwise, we obtain y = 3 or y = 5.

If $3 = 2^x \rightarrow x = \log_2 3 = \frac{\log_{10} 3}{\log_{10} 2}$
If $5 = 2^x \rightarrow x = \frac{\log_{10} 5}{\log_{10} 2}$.

The sum of the roots is $\frac{\log_{10} 3}{\log_{10} 2} + \frac{\log_{10} 5}{\log_{10} 2} = \frac{\log_{10}(3*5)}{\log_{10} 2} = \frac{\log_{10} 15}{\log_{10} 2}$

Question 27: A
Recall the discriminant condition for the existence of real and distinct roots, $b^2 - 4ac > 0$
Using the coefficients in our question, this is: $(a - 2)^2 > 4a(-2)$
$a^2 + 4a + 4 > 0$
$(a + 2)^2 > 0$
Since this is a squared number, all values but a = -2 will satisfy this equation.

ECAA MOCK PAPESR ANSWERS

Question 28: B
We can use the inclusion-exclusion principle to find the probability that none of the balls are red. Since there are 2n blue balls, n red balls, and 3n balls altogether, the probability of drawing no red balls within the two draws is: $\frac{2n}{3n} \times \frac{(2n-1)}{(3n-1)} = \frac{4n-2}{3(3n-1)}$

Therefore, the probability of drawing at least one red ball is equal to:
$1 - \frac{4n-2}{3(3n-1)} = \frac{3(3n-1)-(4n-2)}{3(3n-1)} = \frac{9n-3-4n+2}{3(3n-1)} = \frac{5n-1}{3(3n-1)}$

Question 29: B
The numerator of $\frac{x^2-16}{x^2-4x}$ is in the form $a^2 - b^2$, which means that it can be expressed as the quantity $(a+b)(a-b) = (x+4)(x-4)$

In turn, the numerator can be simplified into: $x(x-4)$.

$\frac{x^2-16}{x^2-4x}$ can therefore be expressed as: $\frac{(x+4)(x-4)}{x(x-4)}$

Which simplifies to: $\frac{(x+4)}{x}$

Question 30: B
At the stationary point, $\frac{dy}{dx} = 0$. Using the product rule: $\frac{dy}{dx} = x^2 e^x + e^x \times 2x$

When $\frac{dy}{dx} = 0$, $x^2 e^x + e^x \times 2x = 0$

Hence, $xe^x(x+2) = 0$

Which shows that the x-coordinates passing through the stationary points of $y = x^2 e^x$ are x=0 and x= -2 respectively. Therefore, the equation of the quadratic function is: $x(x+2) = x^2 + 2x$.

Question 31: B
First, we set the two equations equal to one another: $k(x+4) = 8 - 4x - 2x^2$
$2x^2 + kx + 4x + 4k - 8 = 0$
$2x^2 + (k+4)x + 4(k-2) = 0$
Subsequently, we set $b^2 - 4ac = 0$, as follows: $(k+4)^2 - 4 \times 2 \times 4(k-2) = 0$
$k^2 - 24k + 80 = 0$
Solving this equation yields: $k = 4, k = 20$

Question 32: A
The expansion of $y_1 = (1-x)^6 = 1 - 6x + 15x^2$
The expansion of $y_2 = (1+2x)^6 = 1 + 12x + 60x^2$
The ratio of the second coefficient of y_1 to the third coefficient of y_2 is $-\frac{6}{60} = -\frac{1}{10}$.

Question 33: C
These integers form an arithmetic progression with 300 terms, where n = 300, $a_1 = 1$, and $a_n = 300$. If you substitute these values into the formula for the sum of a finite arithmetic sequence, you will get:

$S_n = 1 + 2 + 3 + 4 + 5 + \cdots + 300$
$S_n = \frac{n}{2}(a_1 + a_n)$
$S_n = \frac{300}{2}(1 + 300)$
$S_n = 150(301) = 45150$

Question 34: D

Recall the double angle formula for sine: $\sin 2\theta = 2\sin\theta\cos\theta$

Since $\sin 2\theta = \frac{2}{5}$, $2\sin\theta\cos\theta = \frac{2}{5}$, $\sin\theta\cos\theta = \frac{1}{5}$

$\frac{1}{\sin\theta\cos\theta} = \left(\frac{1}{\frac{1}{5}}\right) = 5$

Question 35: D

$-11 + 4\lfloor n \rfloor = 5$

$4\lfloor n \rfloor = 16$

$\lfloor n \rfloor = 4$

Since 4 is the greatest integer less than or equal to n, n must be on the interval $4 \leq n < 5$.

Question 36: A

If $a \emptyset b = \frac{a/2}{b}$, then $10 \emptyset n = \frac{10/2}{n} = \frac{5}{n}$ - Equation 1

$n \emptyset \frac{1}{10} = \frac{(n/2)}{\frac{1}{10}} = \frac{5n}{1} = 5n$ - Equation 2

Setting Equation 1 and Equation 2 equal to one another, $5n = \frac{5}{n}$

Thus, $5n^2 = 5, n^2 = 1$, and $n = \pm 1$.

Question 37: D

Since -1 is a zero of the function, $(x + 1)$ is a factor of the overall polynomial. By long division or synthetic division, we can determine that $\frac{2x^3+3x^2-20x-21}{x+1} = 2x^2 + x - 21$.

Factoring $2x^2 + x - 21 = 0$, we get: $(2x + 7)(x - 3) = 0$

The roots are $x = -\frac{7}{2}$ or $x = 3$.

END OF SECTION.

Section 2

Introduction:

- Define "value". Provide a provisional definition, describing value in terms of the benefit that is generated for consumers by a particular product. In this passage, we described the phenomenon of 'price framing', which is the practice of setting prices such that the consumer perceives something as valuable as a result of the price that is offered, although this value need not directly correspond to an objective dollar valuation, as we have seen in the case of the crafty seller.

- The key question: To what extent does price reflect value to a consumer, should perception of value in and of itself be taken as value in and of itself? Who is qualified to speak of 'value' and how can this be regulated, if at all?

Paragraph 1:

- You can suggest that pricing reflects value insofar as it corresponds to a dollar sacrifice that is made in order to obtain a particular product. Prices in a free market are set by the interactions of buyers and sellers, who set prices on the basis of their business requirements – Insofar as we are consumers, we do not actually know what a business's private valuation of the good they are selling is, and therefore we can leave reason to things that reason deserves. Even if we were to take issue with a particular discount, this is the word of the consumer against the word of the seller, and the seller privately knows the price that they wanted to sell at, unless a specific rule for pricing was determined beforehand, it was known, and it was clear that it was deviated from.

- Passage Example:

Price may reflect value insofar as it corresponds to a dollar sacrifice that a consumer makes in order to purchase or obtain a product. Economic theory suggests that if the price were set wrongly, then people would simply not pay. On some level, consumers 'get what they pay for', which suggests that things are priced fairly, but it depends on how you define 'value', because no matter what the promotion was set at, if the consumer paid, then there is no problem.

Suppose we define value as what consumers are willing to pay. If we abide by what we observe from consumer behaviour in response to misleading pricing, there are two possible explanations, one of which is that consumers are irrational and unable to distinguish value accurately or with certainty save for through heuristics and hence are easily deceived, and the second of which is that consumers place a valuation not just on the product in and of itself, but on the nature of price cut displays or the limited time promotion, which can provide the psychological reassurance that the consumer is getting a good deal or affect how much they value the deal, which is priced into their decision when considering different alternatives.

In this line of argument, perception equals value, and the good is priced correctly. Government need not do anything.

Paragraph 2:

- You can suggest that price can be used deceptively, as in the examples that have been given in the text. It is completely possible that businesses can use prices in order to deceive, mislead, to suggest that something is worth more than it actually is, or to imply that something has a value that it never had – But subsequently note that that is based on a specific definition or designation of value, which may not in fact have an objective basis.

- Passage Example:

Suppose that we define value in absolute dollar terms, and suggest therefore that there is an absolute valuation for a good that consumers misperceive. Behavioural economics suggests that humans are not completely rational in determining price, and pricing strategies play into the systematic biases that they have, suggesting that humans can systematically misperceive the absolute value of a product if they fail to consider their biases. For example, suppose the example of a single product that is marketed differently, one presenting the benefits of the good alone, and the other presenting the potential loss that a customer might sustain if they did not purchase that good. Kahneman and Tversky write that human beings exhibit loss aversion, which suggests that if they had the opportunity to pass up what has been presented as a good discount, they may consider it to be more favourable relative to the scenario in which they had been presented with a good deal, even though there was a true absolute valuation, and the consumer was wrong on every count.

- Point out that the role of government, if viewed as protecting consumers, suggests that if this view is held, then the government should regulate.

Paragraph 3:
- To what extent regulating price-setting makes sense, and some possible regulatory schemes.

- Passage Example:

How governments treat 'value' affects how they in turn treat the question of whether they should regulate at all, that is, if you even consider the free decisions of consumers to purchase and sellers to set prices to be a domain in which the government should interfere in the first place. Supposing you do, then your decision may be moderated by whether you believe value is solely determined by perception of the consumer, or there is an objective price for a specific good that should be arrived at independently by different individuals, free markets should theoretically allow for people to set price in accordance to their desires, and to have businesses live or die depending on whether they set the price too high and subsequently receive no customers, or set it too low.

- Make some possible arguments for regulating the information that sellers provide in their prices, and outline their implications. What would happen if we implemented a 'no reference pricing' policy? Would outcomes be fairer, less fair, would they be better? According to what dimension? If 'value' is simply defined as what consumers are willing to pay in dollars, what is better and what is worse? If it is defined as something that is absolute but that consumers routinely misperceive, what is the implication?

Conclusion

Summarize the main points:
➢ How you treat 'value' affects the particular way you consider a particular price.
➢ You may consider 'value' to reflect just the perception of the consumer, in which case there may not be a problem.
➢ On the other hand, you may consider there to be an objective 'value' that consumers misperceive, in which case then perhaps there is an issue.
➢ How you treat value affects how you will regulate, if at all you believe that regulation should be implemented.
➢ Zoom out and say why the question really matters.
➢ Price reflects value, but whether this value is solely contingent on perception or showcases something absolute, is something that is up in the air.
➢ This should affect our decisions and views concerning whether to regulate or not to regulate accordingly.

ECAA MOCK PAPESR ANSWERS

Mock Paper B ANSWERS

Section 1A

Question 1: C
Joseph does not have blue cubic blocks, since all his blue block are cylindrical.

Question 2: C
130°. Each hour is 1/12 of a complete turn, equalling 30°. The smaller angle between 4 and 8 on the clock face is 4 gaps, therefore 120°. In addition, there is 1/3 of the distance between 3 and 4 still to turn, so an additional 10° must be added on to account for that.

Question 3: B
The chance of red is 2/6 = 1/3. To get no reds at all, it must be non-red for each of three independent rolls. The probability of this is $(2/3)^3 = 8/27$. Therefore the probability of at least one red is $1 - 8/27 = \underline{19/27}$

Question 4: D
These three furniture items are compatible with having 6 legs. All the other statements are false.

Question 5: D
Work this out by time. The friends are closing on each other at a total of 6mph overall, therefore the 42 miles take 7 hours. In seven hours, the pigeon, flying at 18moh covers 18 x 7 = 126 miles.

Question 6: E
The passage does not make any supported claims about fruit juice. It gives rationale for both benefits and risks of fruit juice consumption without reaching a conclusion.

Question 7: B
Calculate the overall cost of three stationery sets, then subtract any items not bought. For each item shared between two people, there is one of that item not required. The overall cost is £6.00 per person, £18.00 overall. Subtract one geometry set (£3), one paper pad (£1) and one pencil (50p) to give £13.50 overall cost.

Question 8: C
Argument C is the most convincing. It gives a strong rationale as to why the notion that people should only pay for services which they personally use is likely to have serious adverse consequences on the nation as a whole. Therefore this flawed logic is not suitable to apply to the arts funding dilemma.

Question 9: C
Moving matches 1 and 4 to form a cross inside one of the other cubes will solve the problem. Two squares are broken (the top left hand corner and the overall large square) but four new small ones are created, bringing the total up to seven.

Question 10: D

The white square is opposite brown, since both are adjacent to blue on opposite sides. White and brown cannot be adjacent to each other since the position of the opposite black and red sides makes that impossible.

Question 11: C

We take the overall price to the UK and subtract money which does not go to the farmers. 36,000,000kg at 300p/kg gives £108m. Subtract commission 108 x 0.8, then take 10% of the remaining proceeds as the farmers' share, giving £8.64m

Question 12: B

The first paragraph tells us annual road deaths have fallen, so B is true. The others are false.

Question 13: D

Regression to the mean is a phenomenon observed when a value is variable within a probability distribution. Sometimes by chance it will be at the high or low end, but thereafter it is likely to be closer to what is expected. This can explain the fall in drink driving deaths after the new campaign.

Question 14: E

None of the responses can be reliably deduced from the statement regarding the 'flu.

Question 15: C

Catherine must choose four socks. If choosing three or fewer, it is possible that they could each be of different colour. When choosing four, it is certain that at least two socks will make a matching pair, but possible that there will be two pairs.

Question 16: D

This is another simultaneous equations question. Solve to find x, the normal rate of pay.

$100x + 20y = 2000$ » $60y = 6000 - 300x$ (substitute this)

$80x + 60y = 2700$

» $80x + (6000 - 300x) = 2700$

» $220x = 3300$

$x = 15$

Question 17: D

The easiest way to do this is via simultaneous equations. Let A be the distance travelled by the Plymouth train and B the distance travelled by the Manchester train. Thus:

A = 90x + 45 and B = 70x

The collision will occur when the total distance travelled by both trains is = 405

i.e. A + B =405

Therefore, 90x + 45 + 70x = 405

X =2.25 hours. Thus, collision happens at 12:45.

Substitute x=2.25 into the first equation to give the distance from Plymouth:

A = 90 x 2.25 + 45 = 247.5

Question 18: E

Statement E is not true, the others are true. A pregnant rabbit requires 50 pieces per day and a normal rabbit requires 25. Therefore three pregnant and ten normal rabbits require only 400 pieces per day, not 450.

Question 19: B

If memory of names uses a different part of the brain, then conclusions drawn from this experiment may have no validity.

Question 20: D

Michael pays £60 and £110 = £170 for the painting. He sells it for £90 and £130 = £220. Thus, he makes a profit of £220 - £170 = £50.

Question 21: D

The principle problem is that it does not compare the relative effectiveness of pesticides and natural predators. It might be that pesticides are far more effective at controlling pests, despite the unnecessary excess killing.

Question 22: D

Proportionately, there would be 172 members. Therefore there is an excess of 298 – 172 = 126 members.

END OF SECTION

ECAA MOCK PAPESR ANSWERS

Section 1B

Question 23: A
We can find the common ratio of the series by dividing the second term of the series by the first, yielding the common ratio $r = \left(-\frac{1}{2}\right)x$

Since we know that the fifth coefficient is equivalent to $\frac{1}{32}$, we can solve for the value of x, the first term in the series, by equating 1/32 to the formula for the fifth term of a geometric series:

$\frac{1}{32} = ar^4$

$\frac{1}{32} = x\left(\left(-\frac{1}{2}\right)x\right)^4$

$\frac{1}{32} = \left(\frac{1}{16}\right)x^5$

$x^5 = \left(\frac{16}{32}\right)$

$x = \frac{(16)^{\left(\frac{1}{5}\right)}}{2}$

This is an infinite geometric series with a first term of $a = x = \frac{(16)^{\left(\frac{1}{5}\right)}}{2}$. We can simply find the common ratio by substituting $r = \left(-\frac{1}{2}\right)x = \left(-\frac{1}{2}\right)\frac{(16)^{\left(\frac{1}{5}\right)}}{2}$.

The sum to infinity of a geometric series is given by $S_r = \frac{a}{1-r}$. Therefore, the sum of the series is given by:

$S_r = \dfrac{\left(\frac{16^{\frac{1}{5}}}{2}\right)}{1-\left(-\frac{1}{2}\right)\left(\frac{(16)^{\left(\frac{1}{5}\right)}}{2}\right)}$

$S_r = \dfrac{16^{\frac{1}{5}}}{2+\frac{(16^{\frac{1}{5}})}{2}}$

Question 24: D
$\log_2 3 \times \frac{\log_2 4}{\log_2 3} \times \frac{\log_2 5}{\log_2 4} ... \frac{\log_2(n+1)}{\log_2 n} \leq 10$

Solving the above equation, we have that $\log_2(n+1) \leq 10$. Consequently, $n+1 \leq 1024$. The largest value of n that satisfies this equation is 1023.

Question 25: B
We have:
$(a+b+c)^2 = a^2 + b^2 + c^2 + 2(ab+bc+ca) = 364 + 2(ab+bc+ca) = 26^2 = 676$
so $ab+bc+ca = 156$.
Since b and c are the second and third terms of a geometric progression respectively, let us denote $b = ar$, and $c = ar^2$
We have $a+b+c = a+ar+ar^2 = 26$ and $ab+bc+ca = a^2r + a^2r^3 + a^2r^2 = 156$
$a(1+r+r^2) = 26$ and $a^2r(1+r+r^2) = 156 = 6 \cdot 26$.
We can divide both equations to get
$a^2r(1+r+r2)/a(1+r+r^2) = 6$, or $ar = b = 6$.

Question 26: A
$f(x)$ is a parabola, which is opened up (since its leading coefficient is $a^2 + 1 > 0$), so it has only one extremum and it is a global minimum. $f'(x) = 0 \iff 2(a^2 + 1)x - 2a = 0, \text{ or } x = \frac{a}{a^2+1}$. Luckily for us, $\frac{a}{a^2+1} = \frac{1}{2} \times \frac{2a}{a^2+1} \leq 1/2$ (since $0 \leq \frac{2a}{a^2+1} \leq 1$ for any positive a).

As a result, the minimum in the interval is reached for $x = \frac{a}{a^2+1}$.

We substitute into $f(x)$ to reach

$fmin(x) = f\left(\frac{a}{a^2+1}\right) = (a^2+1) \cdot \left(\frac{a}{a^2+1}\right)^2 - 2a \times \frac{a}{a^2+1} + 10$
$= \frac{a^2}{a^2+1} - \frac{2a^2}{a^2+1} + 10 = 10 - \frac{a^2}{a^2+1} = \frac{9a^2+10}{a^2+1}$

We want this value to be equal to $\frac{451}{50}$.

$\frac{9a^2+10}{a^2+1} = \frac{451}{50}$, so we cross multiply: $450a^2 + 500 = 451a^2 + 451$, or $a^2 = 49$.
Which means that a=7, since $a>0$.

Question 27: B
We know that rain and snow are independent events. If the probability that it will rain is $\frac{2}{3}$ and the probability that it will both rain and snow the following day is $\frac{1}{5}$, we can find the probability that it will snow the day after tomorrow by simply solving the equation:

$\frac{2}{3}x = \frac{1}{5}$
Which yields:
$x = \frac{3}{10}$

Question 28: A
Let us use the double angle formula, $\cos 2\theta = \cos^2\theta - \sin^2\theta$.
Given we know that $\cos 2\theta = \frac{3}{4} = \cos^2\theta - \sin^2\theta$, we know that $\frac{1}{\cos^2\theta - \sin^2\theta} = \frac{1}{\frac{3}{4}} = \frac{4}{3}$.

Question 29: B
If you draw the graphs, you will notice that the two graphs are the reflections of one another in the y-axis.

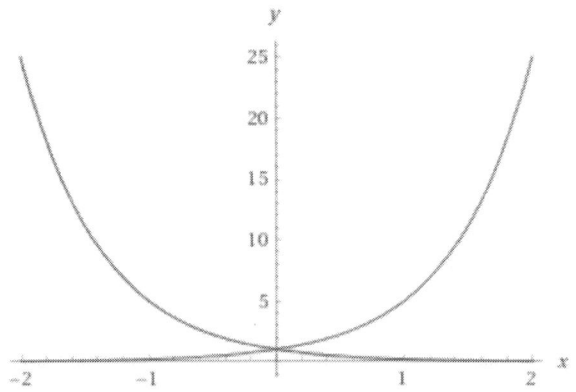

Question 30: C
Note that $1 = \log_4(4)$.
$\log_4(2x+3) + \log_4(2x+15) - \log_4(4) = \log_4(14x+5)$
$\log_4(2x+3)(2x+15) = \log_4 4(14x+5)$
$(2x+3)(2x+15) = 56x + 20$
$4x^2 + 36x + 45 = 56x + 20$
$4x^2 - 20x + 25 = 0$
By factoring,
$4x^2 - 20x + 25 = 0$
$(2x-5)^2 = 0$
Hence, $x = \frac{5}{2}$

Question 31: A

The gradient of the curve is $\frac{dy}{dx} = 2e^{2x-5}$. We know that the gradient of the normal to the curve is $-\frac{1}{\frac{dy}{dx}}$.

Consequently, the equation of the normal is $y - e^{-1} = -\frac{e}{2}(x-2)$.

At the point A, where y=0, $x = 2 + \left(\frac{2}{e^2}\right)$

At point B, where x = 0, $y = e + \frac{1}{e} = \frac{e^2+1}{e}$

Since the area of a triangle is $\frac{1}{2} \times Base \times Height$, the area of the triangle OAB is:

$Area = \frac{1}{2} \times \frac{e^2+1}{e} \times 2 \times \frac{1+e^2}{e^2} = \frac{(e^2+1)^2}{e^3}$

Question 32: D

We know that (sec x + tan x)(sec x − tan x) = $\sec^2 x - \tan^2 x$.
Using the trigonometric identity $\sec^2 x - \tan^2 x = 1$, as well as the information provided in the question, we know that:

$-5(\sec x + \tan x) = 1$

Therefore,

$(\sec x + \tan x) = -\frac{1}{5}$

By substitution, we know that $\sec x - \tan x + (\sec x + \tan x) = -5 + \left(-\frac{1}{5}\right)$

$2 \sec x = -5.2$

$\sec x = -\frac{5.2}{2} = -2.6 = -\frac{13}{5}$

Since $\sec x = \frac{1}{\cos x}$,

$\cos x = \frac{1}{\sec x} = -\frac{5}{13}$

Question 33: B

First, let us find the points along which any potential intersection between the line and the curve would take place, by setting the two equations equal to one another.

$x^2 + (3k-4)x + 13 = 2x + k$
$x^2 + 3kx - 6x + 13 - k = 0$
$x^2 + 3(k-2)x + 13 - k = 0$

Since the line and the curve do not intersect, we know that there must not be any real roots.
As such, by the discriminant condition, we know that $b^2 - 4ac < 0$.

Therefore:

$(3(k-2))^2 - 4(13-k) < 0$
$9(k^2 - 4k + 4) - 52 + 4k < 0$
$9k^2 - 32k - 16 < 0$
$(9k+4)(k-4)$

We know that the critical values therefore extend from $-\frac{4}{9} < k < 4$.

ECAA MOCK PAPESR — ANSWERS

Question 34: A
The distance AC (equivalent to the radius of the circle) can be determined given the coordinates of A and C:
A = (-2,1) C = (5,-3)
Therefore $AC = \sqrt{(5+2)^2 + (1+3)^2} = \sqrt{65}$

To find the length of the line CT, we use Pythagoras' Theorem:
$CT^2 = AT^2 + AC^2$
$CT^2 = 4^2 + 65$
$CT^2 = 81$
$CT = 9$

Question 35: E
$= (6\sin x)(3\sin x) - (9\cos x)(-2\cos x)$
$= 18\sin^2 x + 18\cos^2 x$
$= 18(\sin^2 x + \cos^2 x)$
$= 18$

Question 36: C
Define half the length of the inner equilateral triangle as x, and form a right-angled triangle by drawing a line from the centre of the inner circle to the inner triangle, defining the distance of that line as y.

$\tan 30 = \frac{r}{x}$
$x = \frac{r}{\frac{1}{\sqrt{3}}} = \sqrt{3}r$
$\sin 30 = \frac{r}{y}$
$y = \frac{r}{1/2} = 2r$

Using the formula for the area of a triangle, $Area = \frac{1}{2}ab\sin C$ in conjunction with the formula for area of a circle, $Area = \pi r^2$, we know that:
$Area\ of\ the\ small\ circle = \pi r^2$
$Area\ of\ the\ big\ circle = \pi(2r)^2 = 4\pi r^2$
$Area\ of\ the\ small\ triangle = \frac{1}{2}(2\sqrt{3}r)(2\sqrt{3}r)\sin 60 = 12\sqrt{3}\,r^2$
Therefore, the shaded area is: $Shaded\ area = (12\sqrt{3} - 4\pi + 3\sqrt{3} - \pi)r^2$
$= (15\sqrt{3} - 5\pi)r^2$
$= 5r^2(3\sqrt{3} - \pi)$

Question 37: A
$(3.12)^5 = (3 + 0.12)^5 = (3(1 + 0.04))^5 = 3^5(1 + 0.04)^5$
$= 3^5(1 + 5(0.04) + 5\left(\frac{4}{2}\right)(0.04)^2 + \frac{5(4)(3)(0.04)^3}{6} + \cdots$
$= 3^5(1 + 0.20 + 0.016 + 0.00064)$
$3^5 \times 0.00064 = 0.16$
Therefore, I must obtain 4 terms in the expansion.

END OF SECTION

Section 2

Introduction:
- Raise the point about how art and culture are hotly debated in the national sphere, and about the possible reasons as to why they should or should not receive funding – Or, in the case of austerity, lesser budget reductions amongst a host of various possible alternatives. Consider your case in light of the possible economic implications of reductions in funding for arts and culture, and other points of view as well.

- The key question: To what extent is art and culture something that deserves public funding, and by whom and for what purpose? What are the potential benefits of funding the arts, and what might the potential tradeoffs be?

Paragraph 1:
- Highlight the economic argument for funding the arts, with a specific example that is of relevance.

- Passage Example:
A possible reason to support art and culture is that these are economically beneficial industries to a certain extent. The economic impact of arts funding may be direct, insofar as artists who otherwise would not have been able to perform the tasks that allowed them to generate valuable works, or it may be indirect, insofar as these works may have in turn stimulated ideas within others, causing them to create things of value to the economy. Some economic developments may be contingent on arts and culture, or specifically the individuals who are involved in art and culture who would not otherwise be able to take part in a life of art. Art and culture represents a host of ideas and cultural expressions that might not otherwise exist if not for art and culture funding: To the extent that art is considered valuable to individuals and can facilitate the transfer of ideas within a society, one might consider the provision of art and culture funding to be a public good or an investment.
- Example: Artists in Italy after World War II – Italy supported their designers and artists, and the nation was able to bolster its economy by exporting their products, such as Italian leather, etc, which lends credence to the idea that we should support people with the ideas but not the means. National arts grants, that allow artists to make a living.

Paragraph 2:
- Consider including a counterargument – Although the arts can be economically beneficial, it is hard to assess the impact of funding the arts relative to other industries, and this should be factored into the decision of whether to fund or not to fund.

- Passage Example:
Though we see that the arts account for __% of GDP in the current scenario, the impact of retracting arts funding can be hard to measure, and it is questionable who exactly it will impact, in what scenarios. In a situation in which a government must consider questions of what it will cut from the budget, it must think about what is relatively more important to preserve. Other examples: Example of say, a scenario in which a government must debate between cutting pensions vs. cutting art, followed by justifications concerning why it is that one scenario might have more clear cut benefits than the other.

Paragraph 3:

- To question the approach of making the decision of whether to fund the arts or not on the basis of economic considerations, and what the implications of several possible approaches may be.

- Passage Example:

- On the one hand, making the decision to fund or not fund the arts on the basis of economic considerations may potentially be problematic altogether. If a decision to fund art is made on the basis of economic considerations alone, then a government risks funding only very specific forms of art, not others, which in turn may be justification for governments not to focus on the absolute amount of funding that they provide to art and culture in general, but to zero in on the distribution of those funds in particular. On the other hand, perhaps an economic approach to decision-making on behalf of funding the arts is inappropriate altogether, as arts and culture might be, as Letts might argue, more of a moral consideration than an economic one, as the trade-off might in fact be disproportionate in value relative to what economic impact might otherwise suggest.

Conclusion

- Summarize the points:

➤ Art can be treated as something economically valuable to a country, whether directly or indirectly. These might be considered as justifications to fund or not to fund.

➤ On the other hand, assessing the economic impact of that funding is difficult, and Government should assess the potential tradeoffs that it is making when attempting to assess how and whether to fund.

➤ It may be the case that economic considerations may not be the best way to make a decision on whether to fund or not to fund art, as there may be moral considerations that are involved in that process above and beyond our immediate economic ones.

END OF PAPER

Final Advice

Arrive well rested, well fed and well hydrated

The ECAA is an intensive test, so make sure you're ready for it. Ensure you get a good night's sleep before the exam (there is little point cramming) and don't miss breakfast. If you're taking water into the exam then make sure you've been to the toilet before so you don't have to leave during the exam. Make sure you're well rested and fed in order to be at your best!

Move on

If you're struggling, move on. Every question has equal weighting and there is no negative marking. In the time it takes to answer on hard question, you could gain three times the marks by answering the easier ones. Be smart to score points- especially in the maths section where some questions are far easier than others.

Make Notes on your Essay

You may get asked questions on your essay at the interview. Given that there is sometimes more than four weeks from the ECAA to the interview, it is really important to make short notes on the essay title and your main arguments after the essay. You'll thank yourself after the interview if you do this.

Afterword

Remember that the route to a high score is your approach and practice. Don't fall into the trap that *"you can't prepare for the ECAA"*– this could not be further from the truth. With knowledge of the test, some useful time-saving techniques and plenty of practice you can dramatically boost your score.

Work hard, never give up and do yourself justice.

Good luck!

Acknowledgements

I would like to express my sincerest thanks to the many people who helped make this book possible, especially the Oxbridge Tutors who shared their expertise in compiling the huge number of questions and answers.

Rohan

About UniAdmissions

UniAdmissions is an educational consultancy that specialises in supporting **applications to Medical School and to Oxbridge**.

Every year, we work with hundreds of applicants and schools across the UK. From free resources to our *Ultimate Guide Books* and from intensive courses to bespoke individual tuition – with a team of **300 Expert Tutors** and a proven track record, it's easy to see why UniAdmissions is the **UK's number one admissions company**.

To find out more about our support like intensive **ECAA courses** and **ECAA tuition** check out www.uniadmissions.co.uk/ecaa

Your Free Books

Thanks for purchasing this Ultimate Guide Book. Readers like you have the power to make or break a book – hopefully you found this one useful and informative. If you have time, *UniAdmissions* would love to hear about your experiences with this book.

As thanks for your time we'll send you another ebook from our Ultimate Guide series absolutely <u>FREE</u>!

How to Redeem Your Free Ebook in 3 Easy Steps

1) Find the book you have either on your Amazon purchase history or your email receipt to help find the book on Amazon.

2) On the product page at the Customer Reviews area, click on 'Write a customer review'

Write your review and post it! Copy the review page or take a screen shot of the review you have left.

3) Head over to www.uniadmissions.co.uk/free-book and select your chosen free ebook! You can choose from:
- The Ultimate ECAA Guide – 300 Practice Questions
- ECAA Mock Papers
- ECAA Past Paper Solutions
- The Ultimate Oxbridge Interview Guide
- The Ultimate UCAS Personal Statement Guide

Your ebook will then be emailed to you – it's as simple as that!

Alternatively, you can buy all the above titles at **www.uniadmisions.co.uk/our-books**

ECAA Intensive Course

If you're looking to improve your ECAA score in a short space of time, our **ECAA intensive course** is perfect for you. It's a fully interactive seminar that guides you through both sections of the ECAA.

You are taught by our experienced ECAA experts, who are solicitors or senior Oxbridge tutors who excelled in the ECAA. The aim is to teach you powerful time-saving techniques and strategies to help you succeed for test day.

- Full Day intensive Course
- Copy of our acclaimed book "The Ultimate ECAA Guide"
- Full access to extensive ECAA online resources including:
- 2 complete mock papers
- 300 practice questions
- Online on-demand lecture series
- Past Paper Worked Solutions
- Ongoing Tutor Support until Test date – never be alone again.

Timetable:

- **1030 - 1330:** Section 1A
- **1330 - 1400:** Lunch
- **1400 - 1600:** Section 1B
- **1600 - 1715:** Section 2
- **1715 - 1730:** Summary
- **1730 - 1800:** Questions

The course is normally £195 but you can get **£ 10 off** by using the code "*BKTEN*" at checkout.

www.uniadmissions.co.uk/ecaa-course

£10 VOUCHER:

BKTEN

Oxbridge Interview Course

If you've got an upcoming interview for Oxford or Cambridge school – this is the perfect course for you. You get individual attention throughout the day and are taught by specialist Oxbridge graduates on how to approach these tricky interviews.

- Full Day intensive Course
- Guaranteed Small Groups
- 4 Hours of Small group teaching
- 4 x 30 minute individual Mock Interviews +
- Full written feedback so you can see how to improve
- Ongoing Tutor Support until your interview – never be alone again

Timetable:

- **1000 - 1015:** Registration
- **1015 - 1030:** Talk: Key to interview Success
- **1030 - 1130:** Tutorial: Dealing with Unknown Material
- **1145 - 1245:** 2 x Individual Mock Interviews
- **1245 - 1330:** Lunch
- **1330 - 1430:** Subject Specific Tutorial
- **1445 - 1545:** 2 x Individual Mock Interviews
- **1600 - 1645:** Subject Specific Tutorial
- **1645 - 1730:** Debrief and Finish

The course is normally £395 but you can get £35 off by using the code "*BRK35*" at checkout.

www.uniadmissions.co.uk/oxbridge-interview-course

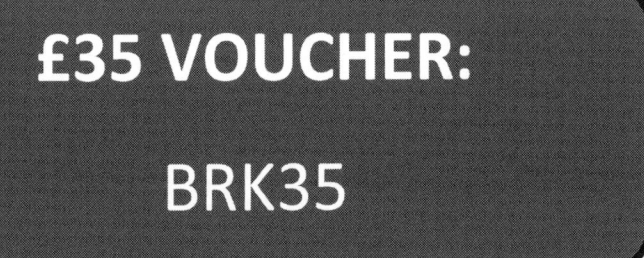

Made in the USA
Columbia, SC
31 August 2018